BADGERED TO DEATH

Published by Canbury Press, 2016
This edition published 2016

Canbury Press,
Kingston upon Thames,
Surrey
www.canburypress.com

Photographs reproduced with kind permission:
Daniel Belton (Badgers)
Andrew Butler (Dorchester)
Emily Lawrence (Dominic Dyer)

Printed in Great Britain by Clays Ltd, St Ives Plc

FSC

ISBN: 978-0-9930407-5-7 (PB)
ISBN: 978-0-9930407-6-4 (EBOOK)

BADGERED TO DEATH

DOMINIC DYER

PRESS

FOREWORD

How viciously fickle we are. We arbitrarily pick and choose which species we like or dislike, normally and sadly based on purely anthropomorphic criteria, and then either laud or loathe them paying scant attention to the realities of their lives, or ours. And once cursed and demonised that tag is almost impossible to redress. Think rat, think fox... damned for historical crimes, firmly fixed as malevolent vermin, even in our supposedly enlightened age. But as this book displays we can also be quick to destroy the reputation of our animal heroes and blight their status with bigotry and ignorance.

For many reasons we had come to love the badger, to cherish and admire it, to protect and celebrate it and of course many still do. But the reputation of this essential member of the UK's ecology has been targeted by a smear campaign which has been swallowed by the gullible and fuelled by those with vested interests. You see, in spite of all the science and all the truths that it outlines, the badger has become a scapegoat. Its been branded a 'bad guy' and is being persecuted as such. It's a terrible shame, but like I said, how fickle, how vicious, how predictably human.

The recent, and as I write, ongoing cull has been deeply divisive and therefore immensely destructive, not only in terms of the piles of dead badgers, but in terms of damaged relationships between the protagonists, and even within their own ranks. Bridges have been burned, reputations ruined and partnerships severed, all with disastrous consequences. As a retort the great service this book achieves is to pronounce the facts; about the animal and its life, about the science that has led to that understanding, about the welfare, moral and economic issues and about the abuse of the scientific truths which should have dictated policies in the first place.

I am a great fan of the author. Like many others I am in awe of his passion, drive and commitment, of his values and motives and of his profound determination to highlight injustice, resist apathy and campaign for proper, information based practices. I know of no one else capable of writing this book, no one who has worked as hard on all fronts to examine and understand the complexities of the issue, no one else who has the bigger picture in such clear sight. And here once again he has done his duty, presenting without ambiguity and bias a precise analysis of this complex affair for the reader to dissect, digest and deliberate upon. And as such it is an immensely valuable resource which provides an opportunity for clarity. And hopefully for change.

In Dominic Dyer the badger has a great and necessary champion, and boy does it need one.

Chris Packham

CONTENTS

Preface: Chris Packham

CONTENTS

'It is clear to me that
the Government's policy
does not make sense.'

———

Lord Robert May,
President of the Royal Society
and Chief Scientific Advisor,
UK Government (1995-2005)

1

A BLACK AND WHITE NIGHT

Thursday 7 May, 2015.

'It's probably the most unpopular policy I'm responsible for'

David Cameron's words were at the front of our minds as we gathered around the table in the Griffin Inn in Witney, his constituency in Oxfordshire. We had fought a long hard campaign against the badger cull, marching in over 30 towns and cities across the country – the largest wildlife protection campaign ever seen in Britain. Now it was the night of the general election.

Everyone knew that the future of the United Kingdom was at stake: everything from the economy to housing, from taxation to immigration. We knew the future of the European badger (*Meles meles*) was too. We wanted a government that wouldn't shoot one of the most loved animals in the British Isles.

The predictions were for a grey night, with no party gaining a majority in the House of Commons. Labour and Conservatives were expected to start scrambling to put together a coalition. But the exit poll at 10pm made clear that David Cameron would be re-entering No 10 Downing Street as the Prime Minister of a majority Conservative government. A policy which made no sense scientifically, which imposed great cruelty on one of the country's surviving large mammals, would resume. There would be another summer bloodbath in the English countryside.

It was the outcome we dreaded. The policy of the Badger Trust was unequivocal: we wanted to see the election of a political party which did not promulgate a badger cull. Since the Badger Trust was a charity, this had caused some problems. On 14 April 2015 Sir Jim Paice, a Conservative former agriculture minister and a champion of the cull, had written to Sir William Shawcross, chairman of the Charity Commission, claiming that the Badger Trust was in breach of the government's new Lobbying Act by planning to support and speak at an anti cull march in Worcester, a key marginal seat, a fortnight before the general election.

We had made the protest in Worcester personal. The slogan was 'Stop Cameron's Cull' – because the cull was part of the Conservative leader's key appeal to voters. Cameron might have been seen as a successful party moderniser, but at heart he remained a traditional Conservative with close links to the farming industry and the shooting and hunting lobby, who grew up as part of a wealthy family in the political and landowning establishment.

He had started his rise in the aftermath of Tony Blair's Labour landslide in 1997, when the Tories retreated to their rural strongholds to rebuild their support. As an aspiring leader, Cameron had harnessed the support of the new Countryside Alliance (CA). The CA had been formed within months of the 1997 election in an attempt to influence the future direction of a weakened Conservative Party on farming and hunting. Cameron also leant on another organisation that wanted to cull a wild animal: the National Farmers Union (NFU). Approaching the 2010 general election, the NFU got what it wanted, when the Conservatives adopted a badger cull policy.

The only problem was that the policy made no scientific sense. It was, factually, a mess. The Conservatives and representatives of landed and agricultural interests wanted to kill badgers to halt the spread of a cattle disease: bovine TB. But there was no support for killing badgers to lower bovine TB from the largest piece of research carried out into the idea, the Randomised Badger Culling Trial (RBCT). Setting aside the moral issue as to whether wild animals should be slaughtered to protect domesticated ones, the RBCT's scientists were asked by the Blair government to establish whether it would be effective to kill badgers. Would a badger cull significantly cut cases of bovine TB, and the growing cost to the taxpayer of compensating farmers forced to cull TB-infected cattle to prevent the disease ripping through their herd?

In essence, the answer was: 'No.'

But the cull went ahead anyway, on the orders of David Cameron. Which was why we were protesting at his constituency at the 2015 general election.

Despite the best efforts of the government to gag me and the Badger Trust, we did reach an agreement with the Charity Commission to remove references to Cameron from the protest march and for me to speak in an individual capacity in Worcester on 25 April 2015 – thus avoiding any infringement of the Lobbying Act.

Worcester was going to be important: an archetypal swing seat. Whoever took it and a handful of other marginals was likely to form the next government. Michael Foster had taken Worcester for Labour in 1997, lost it in 2010 when David Cameron replaced Gordon Brown as Prime Minster and formed a coalition with the Liberal Democrats. A seven per cent swing to Labour would overturn the sitting Conservative MP Robin Walker's majority of 2,982. Every vote would count.

But it was not to be.

As our small band of activists — Emily Lawrence, Laura Paterson, Dave Odell and Gary Hills — drank in the Griffin a fortnight later, we had second thoughts about even bothering to protest, when Sky News broke into live helicopter video footage of David Cameron's car leaving his home for the count. Suddenly we were on our feet, grabbing our badger heads and heading for the cars. We arrived at the count at Windrush Leisure Centre in Witney expecting a heavy police presence and masses of protesters, but it was eerily quiet. A solitary driver was guarding the Prime Minister's election battle bus in a corner of the car park.

He seemed more than happy for us to line up in front of the coach dressed as badgers with our 'Tories Kill Badgers' banner.

At the count itself we found a few protesters against fracking and a very relaxed group of police officers. Expecting to be told we could not get anywhere near the Prime Minister, we prepared our banners and badger outfits on the verge opposite the leisure centre. Then, to our surprise, the police told us we could go right to the line David Cameron would walk past on his way into the hall. Not wasting any time we rushed there, just as the TV helicopter with its piercing lights roared overhead. The Prime Minister's Range Rover and police escort swept in and Cameron was out of the vehicle. As he marched towards the line, all the frustration and anger we felt at the pointless destruction of our badgers boiled over and we started to scream 'Stop Cameron's Cull' with all our might. Visibly shocked, he quickly moved past our protest and into the hall.

Cameron was back as Prime Minister, but we would not be giving up. Only politically did the badger cull make any sense.

2

WOVEN INTO THE LANDSCAPE

Most of us have never seen a live badger; the closest we tend to come is as we speed past a corpse on the side of the road. However these large mammals often live in close proximity to us in the countryside and, more surprisingly, in our towns and cities.

A small but devoted army of human badger-watchers capture a moment of magic at dusk, when they put out small bowls in their back gardens and wait for the badger hours to start.

With its distinctive colouring, and gait, the badger has become one of our most loved animals. *Meles meles* woven itself into our natural history, culture, language, literature, and our place names. Scattered across the landscape are more than a hundred manifestations of the Anglo-Saxon broc, such as Broxbourne (*badger's brook*) in Hertfordshire, or Brockley (*wood where badgers are seen*) in London. When many people think of the wild species of Britain, they turn to the badger.

Badgers are immortalised in Beatrix Potter's *The Tale of Mr Tod* and Kenneth Grahame's *Wind in the Willows*. On the coat of arms of Hufflepuff house at Harry Potter's Hogwarts School of Witchcraft and Wizardry is a badger. The emblem of the national network of Wildlife Trusts is a badger, while three badgers adorn the coat of arms of Tesco, the country's largest retailer.

While there are many examples of a modern fondness for the badger, our relationship with it in the past has been much darker. Indeed the badger has been persecuted by man for hundreds of years. To be 'badgered' means to be pestered and bullied; the mistreatment of the badger has become synonymous with how we treat each other.

As a species the badger had the misfortune to not be considered game or suitable for hunting by the aristocracy and landowning gentry. As such it had none of the status and protection granted to species such as deer or wild fowl. Being a shy nocturnal mammal did not make it an attractive prey to be stalked on horseback or shot from a distance with a rifle or shotgun. Digging out badgers was not the pastime of a gentleman.

Although the aristocracy ignored the badger, this was not the case for farm workers and the rural poor. To them the badger was fair game and badger digging and baiting with dogs was widespread across Britain in the 18th and 19th centuries.

The badger's fate darkened with the arrival of the Industrial Revolution and the coal mining industry. By the early 20th century badger baiting and digging had become a major leisure pursuit for thousands of colliery workers across the country.

This new industrial class of badger persecutor wreaked destruction on badgers and their habitats on an horrendous scale. By the 1950s and 60s tens of thousands of badgers were being killed and hundreds of setts destroyed every year, threatening the very survival of the species in many parts of Britain particularly in mining areas in Durham, Yorkshire, Nottinghamshire and South Wales.

It was during this period that another side of human nature emerged and the first badger protection groups were formed. The first to act was a natural history society in Frodsham, a large village in Cheshire on the outskirts of Liverpool not far from the industrial areas of Burnley, Wigan and Ellesmere Port. Its Badger Group was set up at the start of the 1970s, starting with around 100 members who drove off badger baiters and diggers from setts and physically stood guard over them to protect badgers from persecution. The Gwent Badger Group was formed shortly after, in 1973, and was the main driver for the Wirral and Cheshire Badger Group that was set up in 1979. By the early 1980s badger protection groups has been set up in Yorkshire, Lancashire, Essex, Surrey and Sussex.

Many of the individuals involved in this early badger protection movement lobbied MPs in Westminster to give the species greater legal protection. This resulted in a

Private Members' Bill in Parliament which led to the 1973 Badgers Act, which protected badgers from persecution. (The 1992 Protection of Badgers Act consolidated the legislation and also made it an offence to damage, destroy or obstruct badger burrows, called setts)

However three decades later, despite having some of the strongest legal protection of any species in Britain, the badger is still widely persecuted. The National Wildlife Crime Unit estimates that illegal digs could be killing thousands of badgers in Britain every year.

And cruelty to badgers is now officially sanctioned as part of a policy which 'does not make sense,' according to the country's leading scientist — with the full blessing and power of the British state.

The tragedy for the badger is that just as it received much needed legal protection after hundreds of years of persecution, it stumbled into a new controversy in the livestock and dairy industry in the shape of bovine TB.

3

A DISEASE OF CATTLE

In 1971, something happened that was to prove disastrous for the well-being of *Meles meles*. During an outbreak of bovine tuberculosis (TB for short) in cattle at a farm in the Cotswolds Hills in Gloucestershire, Roger Muirhead, a veterinary officer at the Ministry of Agriculture, Fisheries and Food (Maff), discovered *Mycobacterium bovis* (bovine TB) in a dead badger. Between April 1971 and April 1973, Muirhead and colleagues found 36 of 165 badger carcases were infected with the disease, one in five.

Farmers had found a new enemy in their daily struggle to protect their herds. Bovine TB, which could prove deadly, was the forerunner of a disease that had once regularly killed people in Britain (and still is a killer in parts of the world). A bacterial infection spread by droplets, *Mycobacterium tuberculosis* was a scourge of public health in Britain until the mid-20th century when milk was pasteurised (heated to kill off bugs) and controls at slaughterhouses were tightened up.

Its animal variant, bovine TB (*Mycobacterium bovis*), is spread by the physical manifestations of infected animals: breath, sputum, urine and faeces. It infects a bewilderingly wide range of mammals, including, but not limited to cattle, horses, sheep, pigs, deer, alpaca, llamas, wild boar, badgers, foxes, hedgehogs, domestic cats, mice, and voles.

Bovine TB had been rampant in British cattle in the first half of the 20th century. But from the 1950s, with a law banning the sale of tuberculin-infected milk, a new tuberculin test for cattle, and eradication of whole herds that had been infected, it started to decline. In 1979 the lowest ever prevalence of bovine TB was recorded in the UK. Just 0.49% of herds tested had a TB reactor – which equated to 0.018% of all cattle tested. Significantly, these reductions in bovine TB had been achieved largely though cattle-based measures.

However by then, after the discovery of the diseased badger in Gloucestershire in 1971, Maff had begun to suspect that badgers acted as a reservoir for the disease. While cattle herds with TB could be slaughtered, the ministry figured, badgers carrying the disease could infect new herds. Its examination of 1,934 badger carcasses in the south west between 1971 and 1976 found that 17% of badgers had bovine TB.

Maff's response was to start gassing *Meles meles* using cyanide from 1975 onwards under the Conservation of Wild Creatures and Wild Plants Act. At the same time it established a Consultative Panel on Badgers and TB to provide independent advice and recommend appropriate advice on limiting the transmission of TB from badgers to cattle. The panel included representatives of the NFU, British Veterinary Association (BVA), the Royal Society for the Prevention of Cruelty to Animals (RSPCA) and the National Federation of Badger Groups.

By the late 1970s, the gassing had gone too far for the public. In August 1979, a Maff plan to gas setts at Corndon Down in Devon resulted in a major protest and the formation of the Dartmoor Badger Protection League. Farmers, landowners, retired army officers, civil servants, politicians, academics and even holidaymakers joined the Corndon Down protests, which brought national media attention to the growing public opposition to the gassing of badgers.

Responding to the public mood the government called a halt to badger gassing in 1980 and appointed Lord Zuckerman, President of the Zoological Society of London, to oversee a review of the badger culling and bovine TB reduction policy. Lord Zuckerman reported within 12 months but his findings proved highly controversial. In his report he stated that badgers were 'highly susceptible' to the bovine tubercle bacillus and the resulting infection took a virulent form which encouraged the rapid spread of the disease.

Zuckerman stated:

'There can be no doubt that there has been a significant increase in the incidence of the disease in badgers and that the tuberculosis in badgers many now be spreading from the dense highly infected population in the south west.'

Zuckerman recommended that controlled culling of badgers using cyanide gas re-start, although he did accept that further research was needed into the speed which cyanide gas killed badgers at different concentrations.

However eight months after gassing re-started, the badger received a break from an unexpected source: the Chemical Defence Establishment at Porton Down. It dropped a political bombshell by releasing report which showed that badgers who had been gassed died a lingering, agonising death or suffered brain damage and escaped to other setts, where they could increase the spread of the TB.

Following the publication of the Porton Down report, Peter Walker, the Agriculture Minister, was forced to tell the Commons on 1 July 1982:

'The results of the Porton Down Research imply that there must be some doubt whether all the badgers in a gassed sett die quickly and therefore whether they die humanely.'

Following this statement all further badger gassing was halted but many leading conservationists and scientists remained very unhappy with the Badger Consultative Panel and some of its key members, particularly the British Veterinary Association and the RSPCA, for not calling for a halt to badger gassing far sooner on humaneness grounds.

The review recommended that badger culling be resumed but that on animal welfare grounds gassing should be replaced by trapping and shooting. From 1980 onwards the Ministry of Agriculture, Food and Fisheries pursued a policy of reactive culling, killing around 2,000 badgers a year in high risk TB areas, following TB outbreaks on farms, but only where local veterinary officers believed badgers were implicated. These were small scale reactive culling operations intended to remove just the badgers responsible for transmitting disease.

Surprisingly, none of the badgers killed were tested for TB, meaning there was no way to establish the effectiveness of culling or its impact on lowering bovine TB in the areas concerned.

In 1986 a report by George Dunnet, a professor of natural history at Aberdeen University, called for the urgent development of a live TB test for badgers to prove they had TB before they were killed. In the meantime he recommended that culling should be scaled down so that only badgers in an area where cattle tested positive for TB (termed a 'TB breakdown') be killed.

Over time the NFU and BVA called for this interim strategy to be changed as they believed reactive culling was allowing bovine TB to increase, particularly in the south west, where the damp pastoral landscape provided ideal conditions for both dairy farming and the earthworms and other invertebrates that dominated the badger's diet.

After another sharp rise in the numbers of cattle infected with bovine TB from 2000 onwards (*for explanation of the underlying causes see Chapter 4: New Labour*), pressure again rose for a cull. David Cameron's government introduced one.

One might think that this decision to wipe out in some areas of the countryside a wild animal that had been here for hundreds of thousands of years would be based on sound science. Reasonable – but wrong.

In fact, after 40 years, 35,000 dead badgers, and hundreds of millions spent tackling bovine TB, badger culls ultimately rest on a single piece of research from a poorly conducted experiment by the Ministry of Agriculture in the early 1970s. It took place in January 1975 at the Central Veterinary Laboratories in Weybridge. The field research was designed to prove the cross-transmission of *tubercle bacilli* from badgers to cattle under controlled conditions.

Thirteen badgers were captured from a population known to be carrying TB. Nine of these diseased badgers were confined in a yard measuring 12 metres by 8.5 metres, while the others were held in isolation to replace those that died. Every month the badgers were examined, with blood samples being taken. For the first two years they were also TB tested every two months. Over time three calves were introduced into the confined enclosure with the badgers. Despite being in close proximity to badgers, which were excreting TB, it took six months for the first calf to react positive to a TB test. The second reacted positive at eight months, whilst the third took ten months.

The experiment threw up many problems:
- Badgers made great efforts to escape and had to be held in concrete lined yards with steel doors
- They were highly stressed, which may have increased the risk of spreading TB
- They didn't seem to be especially bothered by the TB. One survived for three and half years after *M.bovis* was isolated from its faeces. Two survived for more than two years

The Ministry of Agriculture scientists who undertook the research accepted that it was inefficient and unnatural and showed that the risk of cattle acquiring TB from badgers was low, even under artificial conditions. In their final research paper they stated:

'In the field the relative risk of cattle acquiring infection from badgers is low and usually only small numbers of cattle become infected. In these experiments the conditions were highly artificial but again the risk appeared to be low.

Calves would regularly exist in the environment for up to five months without acquiring infection even though badgers were demonstrated to be excreting infected faeces during this period. The badgers made latrines in the cattle areas in both experiments but these were left undisturbed by the cattle.'

It was as if the scientists were saying that it was actually difficult for badgers to spread TB to cattle.

Despite this tentative finding, no other field research looking at cattle and badgers and how the disease could spread between the species has been undertaken in mainland Britain in the last 40 years.

So how did the badger cull come about? Here my past furnished some clues. Before becoming deeply involved in the badger protection movement, I had been a high-ranking lobbyist in the food and farming industry. In this former life, as chief executive of the Crop Protection Association, the trade body for manufacturers of pesticides and herbicides, I regularly worked with the NFU President Peter Kendall and his deputy, Meurig Raymond, and maintained close relationships with senior figures across the food chain, including ministers, policymakers and journalists. I knew how political deals that maximised profits for my paymasters were cut, and I knew the extent of the NFU's influence over policymakers and ministers.

Throughout my career I saw how short term economic and political interests trumped the protection of nature.

4

NEW LABOUR

My first encounter with badger politics in the Labour Party came about at a school in Wimbledon in 1995. I was helping to organise a Young Labour outreach campaign to put Tony Blair, Labour's youthful new leader, in front of first time voters in the run up to the May 1997 general election.

Blair gave a polished presentation on his vision for Britain to hundreds of sixth-formers, who could be crucial to his chances of becoming Prime Minister within two years. The young audience listened politely, but showed little excitement or engagement until it came to questions, when they came alive. At this point the confident new leader of the Labour Party was faced with an avalanche of concerns on issues ranging from fox hunting to badger culling, and vivisection to whaling. For the first time I saw Blair under real pressure to show he recognised the importance of animal welfare and wildlife protection to voters.

Following this and other public forums, discussions took place within the Labour leadership to craft a radical election manifesto for animals, which would commit an incoming Labour government to shut down fur farms, ban fox hunting, and end the testing of cosmetics on animals. This animal manifesto ended up in millions of letter boxes in the 1997 election campaign and was undoubtedly an important factor for many voters in returning Labour to office after 18 years with an 180-seat majority. The manifesto did not include any commitments on badger protection, but it was clear that the concerns of Labour voters on animal welfare and wildlife protection issues would influence the development of policy on bovine TB. Indeed, in the first month of the new Labour administration, on 20 May 1997 the Ministry of Agriculture introduced a moratorium on new badger culls. Badger trapping teams were called off by mobile phone.

Before deciding a long-term policy, Jack Cunningham, Tony Blair's first Minister of Agriculture, had to assess a scientific review on TB in cattle and badgers which had been carried out in the last year of John Major's Conservative government by Professor John Krebs, a leading zoologist at Oxford University. In the report, finally published in December 1997, Krebs wrote:

'Recognising the importance of badgers as a source of infection, over the past two decades, Maff has implemented, in succession, a variety of policies for killing badgers in order to control the disease in cattle.

'However, it is not possible to compare the effectiveness of these different policies; nor is it possible to compare any of them with the impact of not killing badgers at all, because there have been no proper experiments.

'However, the indication is that more severe culling policies involving complete, or near complete, removal of badgers from an area, are more effective at reducing the herd breakdown rate than is less complete removal.

'An attempt to target the control at infected badgers only (the 'live test trial') was unsuccessful because of the low sensitivity of the test for TB in badgers.

'We recommend that Maff should set up an experiment to quantify the impact of culling badgers.'

In essence, Krebs was saying that although the evidence from the previous ad hoc culls was shaky it might be the case that destroying all badgers in an area might reduce the number of cattle infected by bovine TB: there should be a trial to find out. He recommended the creation of an Independent Scientific Group to ensure the rigour of any trial. The Labour government welcomed the report as 'representing the best available scientific advice across this area' and was minded to accept its recommendations. There was strong pressure from animal welfare groups and from voices inside the Labour Party to avoid a culling trial. But Jack Cunningham established the Independent Scientific Group on Cattle TB in February 1998 under the chairmanship of Professor John Bourne, a leading veterinarian. Its role was to design the trial.

A battle continued to rage in the Labour Party and the wider conservation movement on whether culling should take place. Meanwhile, the National Farmers Union increased the pressure on Blair to respond to a rise in TB cases in cattle, particularly in the south west.

On 17 August 1998, the new Agriculture Minister, Nick Brown, who had replaced Cunningham in Blair's first reshuffle, announced the government would press ahead with the Randomised Badger Culling Trial (RBCT). Brown was a former shadow Leader of the House and an ally of Gordon Brown – and far more willing than his predecessor to accept the economic justification for badger culling to protect the interests of the livestock and dairy industry. He was soon addressing NFU conferences stating that the government was determined to halt the spread of TB and was willing to confront conservationists by extending badger culling outside of the boundaries of the test sites identified by scientists.

Starting in 1998, the RBCT was conducted in 30 areas of south west and central England designated as high risk for bovine TB. Each 100 square kilometre area was divided into ten sets. Each of these was divided into three smaller areas (triplets). Each triplet could test one of three different methods:

- One where badgers were repeatedly culled across all accessible land by the cage trap and shoot method, by specially trained civil servants (known as proactive culling)
- One where badgers were culled on a single occasion locally and near farmland where recent TB outbreaks

occurred, again by cage trap and shoot (known as reactive culling)
- One where no culling took place (badger setts were surveyed instead). This last method acted as an experimental control against which the culling areas could be compared

The trial would represent the most substantial and coherent evidence base for the evaluation of badger culling.

However, there were some flaws. The trials were subject to protests and some cages were destroyed and badgers released. Nick Brown soon became a hate figure for some of the more extreme elements of the anti badger cull campaign. In June 2000, *The Daily Telegraph* quoted Brown as saying: 'I get death threats about this but they haven't succeeded yet. I have no sympathy with people who say they want to murder the Agriculture Minister because they refuse to accept wildlife should be killed for any reason.'

Most seriously, perhaps, all culling operations were suspended for a year in 2001 because of a large outbreak of highly infectious foot and mouth disease. Foot and mouth can spread from farm to farm by airborne droplets and soon ripped though the UK from a single case in Essex in February. Six million cows, sheep and pigs were slaughtered to halt the spread of the disease, whose epicentre was in Cumbria.

Large parts of the countryside were shut off for months and the May 2001 general election was postponed. Foot and mouth devastated farming and the wider rural economy, including tourism. By the time the disease was halted in October 2001, the crisis was estimated to have cost the UK taxpayer more than £8 billion. But that is only part of the picture, because the restocking of cattle to replace the huge numbers that had been slaughtered as a result of foot and mouth, brought a new problem in the shape of a wave of bovine TB that was sprayed across the country.

Jim Scudamore, Maff's Chief Vet, made it clear to Nick Brown and Tony Blair that key steps should be put in place before any cattle restocking. He recommended shutting down many cattle markets which had poor biosecurity controls that could allow the spread of diseases, such as foot and mouth and bovine TB, and a rigid TB testing and movement control system for cattle, especially before any restocking took place from the south west of England, a TB hotspot.

It is at this point that the NFU crucially influenced policy; to devastating effect in terms of the future control of bovine TB in the UK. The top priority for the President of the NFU, Ben Gill, was to get the farming industry back up on its feet as soon as possible. Farmers across the country faced economic ruin from the foot and mouth crisis, many had lost their cattle herds as a result of the slaughter policy and without replacement cattle they would go out of business.

The very future of the NFU as the voice of farming was at stake, unless livestock farmers could be put back in business. So Ben Gill put huge pressure on Tony Blair and Nick Brown to override the concerns of the Chief Vet and to allow the rapid restocking of cattle, including many from the south west of England.

As a result over the next 12 months hundreds of thousands of cattle were moved across the country, many from

TB hotspot areas in the south west, particularly Devon and Cornwall, without any TB testing and movement controls. Many of these cattle were moved through markets with poor biosecurity, many of which according to the Chief Vet should have been closed down to prevent further disease outbreaks.

The result was the largest increase in bovine TB in cattle ever recorded in the UK. From 2001 to 2002 the number of cattle slaughtered for TB increased by 300%, from under 5,000 to more than 20,000. By the time effective TB testing and movement control systems had been restored in 2003, the figure was almost 25,000. This was not just a disaster for the livestock industry, it was also industrial pollution of wildlife on an unprecedented scale. It was probably the most catastrophic decision concerning the control of bovine TB in 30 years.

5

GORDON BROWN VETOES A CULL

The foot and mouth crisis brought an end to Nick Brown's cabinet career and to Maff itself. After the general election in June 2001, Tony Blair created the Department of the Environment, Food and Rural Affairs (Defra), which took responsibility for the badger cull.

Its new secretary of state, Margaret Beckett inherited a surge in the number of bovine TB cases in the national herd. Following the restocking of cattle after the foot and mouth outbreak without TB testing and movement controls, the number of cases of bovine TB in cattle leapt from just over 5,000 in 2001 to 28,000 by 2003. Beckett was soon coming under pressure from farmers and vets to extend badger culling beyond the randomised trial. The NFU called for the lifting of the moratorium on gassing to control badgers under the Protection of Badgers Act 1992 and even for the mass gassing of badger populations. At the NFU

Conference in London on 7 February 2000, its President Ben Gill said: 'There is a desperate need for urgent action on bovine TB, which is currently rising by 23% a year. The need for interim measures to combat bovine TB by culling badgers outside of the RBCT areas in the new TB hotspot areas is undeniable.'

Even the Prince of Wales became involved in the debate with a 17-page exchange of letters to Tony Blair on farming issues in 2005, which included strong backing for culling badgers. Prince Charles wrote to the Prime Minister:

'I do urge you to look again at introducing a proper cull of badgers where it is necessary. I, for one, cannot understand how the badger lobby seem not to mind at all the slaughter of thousands of expensive cattle, and yet object to a managed cull of an overpopulation of badgers – to me this is intellectually dishonest.'

(This correspondence only came to the attention of the public in May 2015, as a result of a ten-year legal battle by *The Guardian* under Freedom of Information legislation.)

At Defra, Margaret Beckett, shrewd political operator who understood better than most the fine balance that had to be reached in Defra and the Labour Party on badgers, stuck rigidly to continuing, but not extending, the culling trial. Labour was haemorrhaging support in the wake of the Iraq War and Beckett understood the danger of pushing ahead with a highly unpopular policy that would further test the commitment of Labour supporters. Beckett delayed any difficult decisions on culling badgers until after the 2005 general election – a decision for which she was criticised. In February 2005, more than 300 vets signed an open letter in which they claimed that Defra had allowed the 'statutory eradication of the disease to go backwards alarmingly' and warned that the country faced losing its TB-free status, which could harm meat and dairy exports. (If TB becomes endemic in the national herd the EU and third countries can stop cattle leaving the country for disease control purposes.) After Blair's third election victory in May 2005, a rising star in Labour, David Miliband, became the new Secretary of State. He took over at Defra at a critical moment in the cull debate as the randomised trial came to an end and the Independent Scientific Group prepared its findings.

In February 2006, Miliband told the NFU conference that the government was still prepared to cull badgers to tackle the spread of bovine but the science must justify such a decision. Behind the scenes he made it clear to his officials he had no appetite for pushing forward with killing badgers and he was not willing to risk a judicial review on the issue, in view of a weak scientific case for culling. On 12 July 2006 Defra published the results of a consultation that received 47,000 written responses – 95% of the individual responses were against a badger cull. By June 2007 Miliband got his exit strategy from a culling policy, when Professor Bourne's Independent Science Group completed its review of the Randomised Badger Culling Trial. Between 1998 and 2007, the trial had slaughtered 11,000 badgers – at a cost to the taxpayer of £49 million.

The result? It wasn't worth killing any more badgers.

The trial found little evidence of badgers passing bovine TB via urine – and the number of badgers with late stage TB with a high risk of spreading it to other badgers was a very low 1.65%.

The RBCT relied on the dubious 1975 concrete pen experiment to assert that badgers spread bovine TB to cattle. Even so, Bourne determined that although badgers were a source of cattle TB, culling them could make no meaningful contribution to lowering the spread of the disease and could, under certain circumstances, make matters worse. The trial found that while TB in cattle declined inside the proactive badger cull areas, it had risen in the surrounding two kilometre area. The scientists thought the disruption culling caused to badger colonies was forcing surviving, diseased badgers to relocate, spreading bovine to other badgers and cattle. Startlingly, a cull could end up making the disease worse. This was called the perturbation effect.

Secondly, Bourne concluded that weaknesses in cattle testing regimes and movement controls, contributed significantly to the persistence and spread of bovine TB.

In a letter to David Miliband, Bourne wrote:

'While badgers are clearly a source of cattle TB, careful evaluation of our data indicates that badger culling can make no meaningful contribution to cattle TB control in Britain. Indeed some policies under consideration are likely to make matters worse rather than better. Scientific findings indicate that the rising incidence of disease can be reversed, and geographical spread contained, by rapid application of cattle-based measures alone'.

The largest ever scientific inquiry into the issue concluded that culling badgers would not solve the TB in cattle crisis. Tackling problems in the cattle industry, however, would.

The Chief Scientific Advisor, Sir David King, made a last minute challenge to Bourne's recommendations. Having reviewed the findings of Bourne's Independent Scientific Group, Sir David suggested in a report for David Miliband that a cull might still work.

Giving evidence to the Commons Environment Food and Rural Affairs committee (Efra) on 24 October 2007, Sir David explained that although cattle controls remained essential, more thorough removal of badgers could reduce the overall rate of TB in cattle:

> 'If this is done in large enough areas, if we can reduce perturbation of badgers by using wherever possible natural boundaries, and if we can do this over a sustained period of time, as said in the report [to Miliband], we would expect that the incidence of TB in cattle would be reduced, and we would need to couple this with action on cattle as well.'

Visibly annoyed, Bourne told the committee that Sir David's report was 'hastily written and superficial'. Rosie Woodroffe, an ISG member, complained that it was riddled with 'small mistakes'.

In a follow-up letter to Michael Jack, chairman of the Efra committee, Sir David explained that he disagreed with Bourne on a technical point about the impact of the first and second years of badger culling in the trial. Sir David wrote:

'John and I also differ on the extent of the overall beneficial effect. He considers that it is modest whereas I
consider that it is much more significant and that the
benefits will increase over time with sustained culling.'

Professor Bourne said Sir David's recommendations,
were inconsistent with the findings of the randomised trial
and suggested that he was being influenced by political
and economic interests. Sir David, who had consulted other
scientists about the ISG's findings, denied this.

In an editorial on the dispute, *Nature*, the scientific journal, remarked that 'political factors will ultimately overrule
scientific ones when a government takes a decision in a
contentious field.'

That was prescient.

For whatever the rights or wrongs of Sir David's position,
the minister who was taking the decision, David Miliband,
had made up his mind: there would be no cull. In ruling one
out Miliband had the support of the new Prime Minister,
Gordon Brown, who had succeeded Blair in the summer of
2007. Despite a bounce in the polls when he entered No 10
Downing Street, the global financial collapse had mauled
Brown's reputation for prudence and he knew he needed
every vote to keep his grip on power. He abandoned any
thought of culling badgers to please farmers. Unlike the new
Conservative leader, David Cameron, who had stressed the
threat of bovine TB and need to cull badgers in his speech
to the NFU centenary conference in London earlier on 18
February 2008, Gordon Brown avoided all mention of the
issue in his pre-dinner speech.

In July 2008 Hilary Benn, Labour's fifth and final Secretary of State for the Environment, confirmed to MPs in the House of Commons that there would be no mass cull of badgers in the remainder of the Parliament. He told MPs that the risk of perturbation meant that bovine TB could be exacerbated by a cull. He said 'that while such a cull might work, it might also not work.' He said:

> 'It could end up making the disease worse if it was not sustained over time or delivered effectively - and public opposition, including the unwillingness of some landowners to take part, would render this more difficult. I do not think it would be right to take this risk. Therefore ... our policy will be not to issue any licences to farmers to cull badgers for TB control.'

Instead, the government would spend £20 million developing a vaccine against bovine TB that could be administered to badgers. Benn acknowledged his decision would be met with disappointment and anger by farmers, but insisted there was no quick or easy solution to bovine TB.

The NFU threatened to challenge Benn's decision in the High Court, but already attention was shifting to the changing political landscape. Labour's time in office was about to end.

Labour had toyed with badger culling throughout its 13 years in power. Despite its claims to be the party of animal welfare and wildlife protection it oversaw the killing of 11,000 badgers. It deserved credit for not giving in to the pressure from the NFU to extend the culls beyond the 30 randomised cull areas in south west and central England, but the decision by Nick Brown to allow the restocking of foot and mouth-struck herds without TB testing was a serious blunder. Between 2000 and 2008, the number of cattle in Britain slaughtered because of TB leapt by 380%, from 8,123 to 39,007.

Science underpinned Labour's position on the badger cull policy and the randomised trial it commissioned remains, despite its flaws, the best piece of peer-reviewed scientific research on the role of badgers in the spread of TB to cattle undertaken anywhere in the world.

New Labour might not have been the savour of the badger that many in the conservation movement hoped when Tony Blair swept into Downing Street in May 1997, but it had avoided blaming badgers for bovine TB – which was the right view on the facts.

The Conservatives would take a very different view.

6

CAMERON'S CULL

The randomised trial was a thorough peer-reviewed study into the effectiveness of killing badgers to control bovine TB. And it was clear: culling badgers would make no meaningful difference to reducing TB.

In normal circumstances, both the government and the opposition would have accepted the Independent Scientific Group's findings and developed policy accordingly. But in the run up to the 2010 general election, when every rural vote would count, and with commitments already made to the NFU and Countryside Alliance to deliver a badger cull, the findings were inconvenient to David Cameron. He and Jim Paice, the Conservatives' farming stalwart, needed some political cover to undermine Bourne's recommendations and maintain their commitment to culling as a key rural policy. Their salvation came in the shape of Sir David King's challenge to the Independent Scientific Group's findings. Not surprisingly Jim Paice said that by maintaining that a thorough cull could work the Chief Scientist was merely confirming what many farmers believed: that the randomised trial had not come to the right conclusion about tackling bovine TB.

At the 2010 general election, the Conservative Party promised to undertake a badger cull.

The importance of that commitment for the Conservatives lay in the dark days of the Blair landslide in 1997 that gave Labour a 180-seat majority.

For the first time its modern history the Tories were no longer a power in many towns and cities and their young new leader, William Hague, was leading a rump of a party which was increasingly reliant on rural constituencies for its remaining support. This did not go unnoticed within the farming, landowning and hunting interest groups which had traditionally backed the Tories. To them the disastrous result of the 1997 election was an opportunity to reclaim their power and influence over the future direction of the party on its long trek back to power.

A key development in this process was the creation of the Countryside Alliance, in summer 1997, by the merger of three organisations: the British Field Sports Society, the Countryside Business Group, and the Countryside Movement. Ostensibly its objectives were to help promote and defend the British countryside and rural life in the media and Parliament and to be a voice for people from all walks of life in the countryside. From its earliest days it was clear that it was primarily a political campaign group for the landowning and hunting lobby.

Within a few years of being established the Countryside Alliance had more than 100,000 members, many of whom were Conservative activists, supporters, MPs or peers. Polling showed that on average over 60% of its members supported the Conservatives and the organisation increasingly became the party's rural voice of opposition to Tony Blair's Labour.

Hague attempted to rally the Tories around an increasingly Euro-sceptic message, making the defence of the pound a cornerstone of his leadership. He also tried to project a more youthful image as leader, with trips to theme parks and the Notting Hill Carnival, but he faced an uphill task. Despite his often good performances in the House of Commons, the leadership came too early for him in his career and the party was so severely broken no one could put it back together again to win the next election. On 13 September 2001, Hague resigned after failing to make any inroads against Labour in May 2001.

On 22 September 2002, the Countryside Alliance brought 400,000 supporters onto the streets of central London in one of the biggest demonstrations ever seen in the capital. Many leading conservative politicians, including the new Conservative leader Iain Duncan Smith, joined the march. It was billed as a wake-up call for the Labour Party about the scale of rural opposition to a number of its policies including its plan to ban hunting with hounds and its failure to begin a badger cull to help farmers struggling with rapidly rising levels of bovine TB after the 2001 foot and mouth outbreak.

However the protest was just as much a shot across the bows of the Tory Party who many in the Countryside Alliance believed had failed to realise the electoral importance of harnessing the support of the farming, landowning and hunting community to rebuild its fortunes.

One new Tory MP who needed no convincing of this argument was the new member for Witney in Oxfordshire, David Cameron who entered Parliament in 2001 in the midst of the Tories' second largest ever electoral drubbing.

Many now view Cameron as a successful moderniser of the Conservative Party. He undoubtedly decontaminated the Tory brand during the Blair years and put the party firmly in the political middle ground economically and socially. However on the countryside and issues such as fox hunting and badger culling, he remained very much a traditional conservative rooted to the farming and landowning community, which forms such a strong part of the party's support in rural constituencies.

Prior to becoming leader of the Conservative Party in 2005, he had stalked and shot deer, been a regular participant at game bird shoots and agricultural shows and had ridden with his local fox hunt, the Heythrop, in his Oxfordshire constituency. He fully understood the importance of harnessing the support of organisations such as the Countryside Alliance and National Farmers Union to establish a strong bedrock of support for the Tory Party on its slow return to power.

Cameron admired many of Blair's skills as a political leader but he understood that Labour was increasingly seen by many in the countryside as a party of the towns and cities, out of touch with the concerns of farmers, landowners and field sports enthusiasts. He believed that Labour had made itself electorally vulnerable through its commitment to ban hunting with hounds and its failure to allow a wide-scale badger cull.

When the short painful leadership by Iain Duncan Smith ended with a no confidence vote in November 2003, Michael Howard took over as Conservative leader. Unlike Hague and Duncan Smith, Howard was an experienced politician, who took over at a time when the first big cracks appeared in New Labour project as a result of Tony Blair's backing for the invasion of Iraq.

Cameron had been a special advisor to Howard during the last Conservative government and he quickly brought Cameron into the shadow cabinet. By the 2005 general election, Cameron was head of policy coordination. It was in this post that he started to develop his pro-hunting, pro-badger cull agenda as priorities for the Tories to regain support in rural constituencies. With his close links to the Countryside Alliance and the NFU, and the ear of his party's leader, he was perfectly placed to ensure their priorities were reflected in the policies the Conservative Party put forward in its 2005 manifesto.

By this time the Tories were recovering politically, partly because of the deep divisions which opened up in the Labour Party over Iraq. Michael Howard knew the Tories would not win in 2005, but with Tony Blair's popularity plummeting, he was confident the party could make significant inroads back into its traditional heartlands in both urban and rural seats.

The election of 5 May 2005 resulted in Labour winning a historic third term but there was little to celebrate in Downing Street. The Iraq war had resulted in millions of traditional Labour voters staying at home or defecting to the Liberal Democrats. Labour's vote dropped by 5.5% and it lost 47 seats.

The Tories' share of the vote nudged up by 0.7%, but a well-managed campaign focusing on key target constituencies, many in the south east and London, saw them gain 36 seats. Cameron was given much credit for his role in the campaign and for helping to restore confidence among party members that the Conservatives were at last well on their way to electoral recovery after almost a decade in the political wilderness.

By now the Countryside Alliance and the National Farmers Union were increasingly looking to Cameron as a future leader of the party who would champion their interests. On the day after the 5 May general election, Michael Howard announced he would stand down as leader, with a leadership election to take place in the autumn.

A number of candidates stood for election but the two strongest contenders were David Cameron and David Davis. Davis was a very different kind of politician to Cameron. He grew up on a council estate in Tooting in London and began a career with Tate & Lyle. He entered Parliament long before Cameron in 1987 at the age of 38 and served as Europe Minister in John Major's government. Unlike Cameron,

Davis was not close to the farming industry and had no interest in hunting or shooting. He was a more authentic Thatcherite, a non-establishment figure with a strong business background, a firm belief in civil liberties, and a pragmatic approach to the issue of Europe. However he lacked the charisma and energy of the younger Cameron and the wider support he enjoyed in the grassroots of the Tory Party.

All the leadership candidates were given an opportunity to speak for 20 minutes at the Tory Party conference in October. Davis entered the conference as the favourite for the leadership but his speech was wooden and he showed a lack of confidence and energy. Cameron dumped the autocue and made what many in the hall considered an energetic and inspiring speech. By the time the conference was over Cameron was the favourite to lead the party and he never looked back. On 6 December 2005, David Cameron was declared Conservative Party leader with 68% of the vote to David Davis on 32%.

Cameron knew that if he was to become Prime Minister he would have to go after every vote at the next election. When it came to rural issues and the environment he embarked on a bold two-pronged strategy which involved both support for fox hunting and badger culling but also a wider greening of the party to focus on issues as climate change and sustainable farming. He was gambling on the Tories being able to shore up their support in the more traditional farming and landowning community, while reaching out to the wider electorate with a broader green agenda.

In the early years of his leadership Cameron surrounded himself with advisors such as the climate change champion Lord Gummer and the millionaire environmentalist Zac Goldsmith. Cameron put a windmill on the roof of his London home, promoted the virtues of organic farming, and even made a visit to the North Pole to focus attention on climate change. He could also be found at NFU conferences and Countryside Alliance receptions praising the virtues of intensive agriculture, calling for badger culls, and supporting calls for the return of fox hunting.

By the time of the economic downturn, David Cameron knew that the key focus at the next election would be restoring growth, creating more jobs and cutting the growing deficit. His love affair with the green movement started to wither. There was no more talk of windmills or organic vegetables; the message was now all focused on the economy.

Speaking at the NFU centenary conference in 2008, the NFU President, Peter Kendall, said it was time all political parties were willing to challenge their largely urban supporter base on the need for badger culling despite the controversy it generated.

Cameron's keynote speech at the conference was on global food security and the need for UK to become more self-sufficient in food production for the future at a time of rapidly rising commodity and food prices.

During the debate that followed his speech, he committed an incoming Conservative government to cull badgers in areas of England where bovine TB was a significant problem for the livestock and dairy industry.

To Cameron the cull was not about the danger of losing urban votes, but about regaining support for the Conservative Party in rural communities, which would open the door for him to enter Downing Street. I remember watching him from the audience. In comparison to the Prime Minister Gordon Brown, who addressed the conference in the evening, Cameron was far more confident and upbeat. Unlike Brown, he was clearly amongst friends and felt far more at home with farmers and landowners talking about issues such as badger culling, pesticide policy and GM crops, all of which were becoming contentious for Defra with Hilary Benn at the helm.

The 2008 NFU centenary conference sealed the deal for the farming lobby when it came to who it would be backing at the next election. Cameron had the right the background and commitment to their issues and, with Brown struggling, he had a real opportunity to return the Tories to power after over a decade in the wilderness.

Following the conference Cameron moved quickly to shore up his support within the farming industry on bovine TB and badger culling. He embarked on a tour of farming constituencies making clear commitments to NFU gatherings in the south west that an incoming Tory government would no longer sit on the fence on bovine TB and would push forward with a badger cull despite hostile public opinion.

He also reached out to the Countryside Alliance attending receptions, speaking at their Tory conference fringe events and writing guest articles for their magazine. He spoke fondly of his roots in the countryside and how proud he was of Britain's rural identity and heritage. He also emphasised his strong support for country sports including shooting and hunting and made a clear commitment, should he become Prime Minister, to repeal the ban on hunting with hounds.

Cameron also gave strong support to efforts behind the scenes to get key figures from within the farming and hunting community onto Tory MP selection lists for safe seats ahead of the next general election. This included the selection of the chief executive of the Countryside Alliance, Simon Hart, as the prospective Tory candidate for Carmarthen West and the farmer and former MEP Neil Parish for the seat of Tiverton and Honiton.

By the election of 6 May 2010 Cameron's strategy for locking in the Countryside Alliance and National Farmers Union was complete. He had gained their trust and confidence, made clear commitments on badger culling and repealing the ban on hunting with hounds and supported the selection of their candidates to become MPs.

In the election campaign both the Countryside Alliance and NFU mobilised their members in rural seats to campaign for Tory candidates who were committed to supporting their positions on hunting and badger culling. Countryside Alliance members and supporters campaigned with local candidates, canvassing, handing out promotional leaflets, and making phone calls. Regional events were held by NFU and the Alliance for parliamentary candidates, town hall meetings, and political hustings.

The Conservatives won 36% of the vote, against Labour's 29%. However Cameron still remained 20 seats shorty of a majority, resulting in the first hung Parliament since 1974. Brown stayed on in No 10 hoping to cobble together a coalition with the smaller parties, but he was ten seats short of a majority.

Many in the Tory Party were critical of Cameron for not achieving a stronger victory over Gordon Brown. To be fair, Cameron still had a mountain to climb in 2010 despite all the difficulties Labour faced following the 2008 economic crash. His strategy of focusing on rural issues and working closely with the Countryside Alliance and the NFU was successful. The Tories picked up many rural seats from Labour in the Midlands, East Anglia, south and south west of England. In an election where every vote really did count, the time and investment he made in supporting the farming lobby and its push for badger culling paid dividends at the ballot box. The Conservatives' 306 MPs formed a coalition with the 57 Lib Dem MPs.

On 11 May 2010, David Cameron, walked into No 10 Downing Street, smiling broadly. The fate of the badger was sealed by the time he and the Liberal Democrat leader Nick Clegg held their coalition agreement press conference on 12 May 2010 in the rose garden of Downing Street. Cameron was fortunate in finding many Liberal Democrat MPs had rural seats in south west England where they were under significant pressure from their farming constituents to back a badger cull. When it came to the fine print of the coalition agreement, little time was needed to consider the cull: both Cameron and Clegg were willing to start killing badgers. Buried deep in the coalition agreement of 20 May 2010 were the words:

'As part of a package of measures we will introduce a carefully managed and science-led policy of badger control in areas with high and persistent levels of bovine TB'.

Despite this, relatively little happened in the first year of the coalition. On 19 July 2011 the pace quickened when its first Secretary of State for the Environment, Caroline Spelman, a scientist by training, announced a package of measures to reduce bovine TB, including enhanced testing of cattle herds, more training of farmers on biosecurity – and a badger cull.

She launched a consultation on the method to be used in the cull, which, controversially, would be the cheap option of free shooting. In the randomised trial, the badgers had been trapped in cages and then shot. Although the trapped badgers in the trial still died, they were at least despatched quickly and surely. Under Spelman's plan, marksmen with a deerstalking licence would roam the countryside shooting badgers at night, under licences issued by Natural England. This meant that badgers shot from far away might be wounded and amble off to die a slow and agonising death. Defra put the cost to the government of dealing with bovine TB at £100m a year. Spelman said: 'If we don't change what we're doing, [bovine TB] will cost the country a £1bn over the next 10 years.'

Farmers would pay for the cull. Its target would be to kill 70% of the badgers in each culling zone, meaning the policy could kill up to 30,000 badgers a year. The government estimated that culling badgers would reduce the increase in incidents of bovine TB infections in cattle by up to 16% over nine years.

All that was needed, after a satisfactory consultation, was a final decision from Caroline Spelman. Despite protests from Labour, scientists and naturalists, that came on 14 December 2011, when the Secretary of State told the Commons that the pilots culls would begin in two counties, Somerset and Gloucestershire, in autumn 2012. An independent panel would deem the effectiveness and humaneness of the cull, before a decision was made to extend the trial to four areas over a further eight years.

The Labour government had held out hope that an oral TB vaccine for badgers would be available by 2015, but Spelman told MPs that a vaccine was 'years away' and no-one could be certain when it would be ready.

The Guardian estimated the costs of a nine-year cull. Farmers would pay £56m for surveying, culling and coordination and taxpayers £36m for licensing and monitoring. That came to a total of £92m, before the addition of the cost of the courts or policing protests in the cull zones. Between 23,330 and 35,000 badgers a year would be killed. Over eight years of culling, the death toll of the protected badger would be 70,000-105,000 badgers.

The cull would begin in the autumn of 2012.

Still, the announcement was welcomed by the NFU. Its president, Peter Kendall, said:

'Today is another massive step forward in achieving our end goal of a healthy countryside – both for badgers and for cattle.'

7

WILDLIFE OVER BUSINESS

While the coalition's plans for the badger cull travelled towards their deadly destination, I was myself on a journey away from the corporate world and into a full-time role protecting wildlife, and specifically badgers. Alongside my work as a corporate lobbyist, I was chairing Care for the Wild, a wildlife protection charity, and was well aware of the many threats facing wildlife at home and abroad.

Following a trip to Kenya in 2011 to look at anti poaching work in the Maasai Mara National Park funded by Care for the Wild, I wrote an opinion article in *The Sunday Times* calling for greater military support for African nations to protect their elephants and other endangered wildlife from poachers. Published shortly before the CITES Summit on the trade in endangered animals in Doha, the article provoked a debate among politicians and conservationists. This made me start to think that the time had come to fight for wildlife, rather than for the corporate business world.

The badger cull was making it increasingly difficult for me to continue as chief executive of the Crop Protection Association, which involved maintaining a close working relationship with senior figures across the food chain including the top brass of the NFU. In 2010, I was even approached by head-hunters to see if I might become the next Director General of the NFU. This good working relationship with the NFU could be maintained, but only at the cost of not going public on my concerns about the badger cull. Up to 2011, I managed this balancing act well, but by 2012 the NFU was becoming impatient with the delay in implementing the cull.

Then a chance encounter on a suburban street made up my mind.

My first encounter with a badger had come two decades previously on a beautiful moonlit night on the Isle of Wight, in the cliff-top coastal town of Ventnor. On a weekend with my partner Michelle at a hotel which specialised in badger watching we sat in its conservatory after midnight watching its large gardens, which backed onto woodland. The owners had put out bowls of food in the garden to entice the creatures. One by one the badgers ambled across the lawn. At one point, five were feeding from the bowls. Suddenly, a fox appeared; it was wary of the badgers but in time mustered enough courage to join the midnight feast and cap an enchanting and wonderful spectacle. The joy and excitement of that first encounter with wild badgers had never left me; the black and white faces, the thick glossy coats looking silver in the moonlight, the tufty little ears, dark black eyes and short bushy tails.

In spring 2012, 18 years after I saw those badgers, and with thoughts of working full-time for wildlife swirling round my mind, I was wandering along a street in Kingston upon Thames, Surrey, in the early evening, when suddenly a large badger walked directly towards me. On seeing me blocking its path, the creature rapidly turned tail and dashed across the road. Luckily I was able to signal to an approaching car, who slammed on his brakes and avoided an all too common cause of death for these creatures.

On returning home that evening and explaining my latest encounter to Michelle, she joked it was a sign: the badgers had sent out an emissary to recruit me into defending them against the government and National Farmers Union. I laughed but, in a way, we both knew where this was heading.

I had a choice to make: stay a highly paid corporate lobbyist and remain silent on badger culling, or stand up — and confront the very people I had worked closely with for many years to oppose the killing of badgers. With the support of Michelle and close friends and family, I chose to go with my heart. By July 2012, I was working as full-time policy advisor at Care for the Wild focussing on the badger cull. Care for the Wild had a long track record of standing up for UK wildlife and had worked closely with the Badger Trust for decades in promoting awareness of badger ecology and behaviour and the need to protect the species. It had also campaigned against previous efforts to cull badgers and was preparing to do the same again in response to this latest threat. My bank balance was not looking too healthy, but for the first time in my career I felt able to use my skills and experience to make a real difference in conservation.

I was able, uniquely, to look at the badger cull debate not just as a conservationist, but as someone who had worked at the centre of government and the food and farming industry for over 25 years. Unlike many who came into the badger protection debate, I understood the devastating economic and social impact of bovine TB on farmers and the difficult decisions ministers faced in developing a policy to eradicate the disease. I knew how the NFU worked and the influence it had over Defra policymakers and ministers and I was only too aware of the huge influence the food, farming, fisheries and agro-chemical industries had on the development of farming and environment policy compared to conservation and wildlife protection interests.

On leaving the world of government and corporate lobbying, I was keen to share my thoughts on these issues with a wider audience in the form of a lecture or debate at a science or conservation body. Fortunately, under its 21st Century Challenges programme, the Royal Geographical Society hosted a debate looking at the impact of science and environment policy on the countryside. Chaired by the BBC's Science Correspondent Tom Heap, it was titled: 'Crisis in the Countryside'

The organisers wanted to focus on a number of issues ranging from pesticides and GM crops policy to the badger cull. With my background in government, industry and wildlife conservation I had worked in these all areas and I was asked to join the speaker panel alongside Craig Bennett from Friends of the Earth and Professor Nick Pidgeon from the University of Cardiff.

Over 500 people packed the Ondaatje Theatre on Exhibition Road in Kensington on 26 June 2013. Tom Heap introduced the debate by focusing on the growing public concern over the lack of transparency concerning decision making by government over complex scientific issues such as pesticides controls, GM crops and the role of badgers in the spread of TB to cattle. He told the audience: 'We now have a crisis of confidence in relation to decisions by government which effect the future of our countryside wild-life protection and the farming industry'.

In my speech, I pointed out that all Environment Secretaries struggled with making decisions on environment or farm policy issues in the face of competing arguments from industry, conservation bodies and scientists. However, I said the balance was all too often weighted in favour of industry and short-term political interests, rather than the protection of the environment and wildlife.

It had been horrifying to see the way the National Farmers Union pressurised Tony Blair and his Agriculture Minister Nick Brown to override the concerns of the Defra Chief Vet Jim Scudamore and allow the restocking of cattle after the foot and mouth outbreak in 2001 from the TB hotspots in the south west.

I explained that during my time as chief executive of the Crop Protection Industry between 2008 to 2012, I was left in no doubt that the government and farming indus-try's priority on pesticides was productivity and food prices; which went a long way to explain their resistance to using the precautionary principle when it came to assessing the long-term impact of neonicotinoid insecticides on bees, butterflies and other pollinators.

However, I saved my most savaging criticism for the badger cull policy. In all my time working in government and industry, I told the audience, I had never seen a policy which was being so clearly pursued for short term political and economic interests. To kill thousands of badgers by an inhumane, unproven shooting method without testing any of them for TB to measure the impact on lowering bovine TB in cattle, was not science-based policy. The badger cull showed Defra to be an arm of government which no longer protected the environment, but protected business from the environment.

The reaction from the audience to my views was positive but during the debate Tom Heap pressed me on my objection to badger culling, by drawing attention to his recent experience reporting for Countryfile on how Ireland had been tackling the problem of bovine TB by killing thousands of badgers.

My response was unequivocal: Britain is not Ireland. Since the 1980s when the Irish government started culling tens of thousands of badgers, TB rates in cattle have risen and fallen. Even the Irish government admits 'the difficulty in attributing trends to a single factor and the cyclical nature of the disease.' I stressed that public concern about animal welfare and wildlife protection was much higher in Britain and people would not remain silent while the government and farming industry did backroom deals to remove a protected species from large areas of the country without any real scientific justification. If the government and NFU pressed ahead with its plans to cull badgers, unlike in Ireland we would see a huge upsurge of public anger and protests across the country.

While I uttered these words, I had already decided that I wanted to play a key part in a protest movement that would unite compassionate people in towns and cities across the country.

The cull was coming closer and closer to becoming a bloody reality, while the science against it was becoming ever starker.

8

AWKWARD FACTS

Despite her commitment to removing badgers from the countryside, Caroline Spelman would not get to oversee the cull. She was sacked as Defra Secretary of State in David Cameron's first reshuffle on 4 September 2012, after a difficult period in charge of the department which culminated in her making an apology for a botched attempt to sell 258,000 hectares of state-owned woodland.

Her successor was Owen Paterson, a hard right Eurosceptic with a rural constituency and a strong determination to fight off the climate change campaigners and other environmentalists who he blamed for hampering the development of the rural economy. Paterson's brief was to provide stronger leadership within Defra and to pursue difficult policies on issues such as the badger cull, pesticide use, GM crops and reform of the European Union's common agriculture and fisheries policies.

As a child in Cheshire, Paterson had kept two orphaned badger cubs which had been handed to his family by a local farmer. Paterson named the cubs Bessy and Baz and kept them in the boiler room of the family home — until they escaped to a nearby sett. Despite his fond childhood memories, Paterson was clear in his mind that badgers have to be culled — to tackle bovine TB. 'I am a badger lover,' he explained. 'I have extremely fond memories of Bessy the badger. But what I find inconceivable is that some so-called badger lovers are prepared to let these animals suffer from what is a horrible disease.'

Although he was determined to push forward with a national cull, first he had to deal with a setback. The cull had been intended to start in the summer of 2012. However, because of the likelihood of widespread protests in the culling zones in Gloucestershire and Somerset, the police had requested it be put off to allow the country to stage the Olympics and Paralympics. Then, Paterson told the Commons on 23 October, weeks after his appointment, that 'exceptionally bad weather' had disrupted preparations and badger numbers in the two counties had been higher than expected. The cull contractors could no longer be sure they would kill 70% of badgers in those areas in 2012, so the cull would be delayed until 2013.

Despite having to delay the start of the cull, Paterson made it very clear that he would not be pushed off course by growing public, political and scientific opposition to the policy. Bovine TB had led to the slaughter of 26,000 cattle the previous year, he told MPs, declaring: 'I'm entirely convinced that the badger cull is the right thing to do.'

In his statement to the Commons, he baldly re-stated the science:

> 'Research in this country over the past 15 years has demonstrated that cattle and badgers can transmit the disease to each other; culling badgers can lead to a reduction of the disease in cattle if it is carried out over a large enough area and for a sufficient length of time.'

Leading TB scientists who had spent years studying the issue disagreed vigorously that culling was the answer to TB in cattle. In a letter to *The Observer* on 14 October 2012, days before the culls had been due to begin in Somerset and Gloucestershire, more than 30 animal disease experts described the cull as a 'costly distraction', saying it was too short-term and small-scale to monitor the impact of culling, and could make bovine TB worse. Professors Krebs and Bourne – who led the randomised trial research – both signed the letter, which stated:

> 'We recognise the importance of eradicating bovine TB and agree that this will require tackling the disease in badgers. Unfortunately, culling badgers as planned is very unlikely to contribute to TB eradication.'

Separately, Lord Krebs said:

> 'The scientific case is as clear as can be: this cull is not the answer to TB in cattle. The government is cherry-picking bits of data to support its case.'

Lord Robert May, a former chief scientist and President of the Royal Society, the leading organisation for British scientists, who also signed the letter, warned: 'It is very clear to me that the government's policy does not make sense. I have no sympathy with the decision.' Witheringly, he added:

'They are transmuting evidence-based policy into policy-based evidence.'

The scientists knew that any decline in TB was far more likely to come from testing and movement controls. The evidence that badgers transmit bovine TB to cattle is very patchy, being based on that highly-limited study in 1975. Any infection might be the result of cattle coming into contact with badger faeces or urine in pasture areas or in farm buildings, which badgers might enter to feed on stored cattle feed and maize silage.

There is, though, evidence that badgers mostly try to stay away from cattle. Research carried out by P F J Benham and D M Broom in 1989 and published in the *British Veterinary Journal* looked at the interaction between badgers and cattle on pasture, using both natural and artificial conditions. They found that badgers generally avoided cattle, often keeping 10 to 15 metres away and rapidly fleeing from any approaching group of cows. From these observations they concluded that badgers would not result in direct transmission of TB to cattle via their breath or by direct contact. They also found that only a minority of cows were willing to graze close to badger urine or faeces and that most avoided badger products altogether, making infection by ingestion very unlikely.

This has been backed up by more recent research from zoologists at Trinity College Dublin, who, working with the Department of Agriculture Food, Marine and National Parks in Ireland, followed 50 badgers at night in County Wicklow using GPS tracking technology. Published in 2015, the four-year study captured one of the largest data sets of badger movements ever. Its key finding was that badgers avoided fields and farmyards where cattle were grazing or present. Badgers would roam great distances at night, but rarely came close to a cow. Associate professor in zoology Dr Nicola Marples said the study 'found that badgers clearly avoided fields if cattle were present. If it's a field they like, they will return when the cows are not there.'

Bovine TB has also been identified in a number of wildlife species other than badgers in the UK. Research data collected by the Maff from 1971 to 1996 found TB in species ranging from deer and foxes to hedgehogs and cats. Some were more likely and some less likely to be carrying TB than badgers. In the case of fallow deer (18.5%) and farm cats (16.7%) the prevalence of the disease was higher than in badgers. It was lower in foxes, moles, mink, and rats.

Of the 11,000 badgers killed during the RBCT, 16% had TB – but only 1.65% were suffering from late-stage symptoms with visible lesions and high risk of excreting the disease through their skin or urine to other badgers. Peculiarly in the circumstances, given the vast amount of money being spent on the badger cull, no significant research has been undertaken into the possible impact of

TB in wildlife on cattle other than badgers. This is despite the fact that species such as deer, cats and rats are far more likely to live in close proximity to cattle in fields or farm yards than badgers.

A 2013 paper derived from the RBCT suggested that TB breakdowns caused by badgers was not around 1 in 20 as stated in the randomised trial, but 1 in 27. This means that the total of 472 TB breakdowns in the randomised trial, only 18 may have been caused by badgers. So if badgers are passing TB to cattle it is very rare. The vast majority of cattle infected with bovine TB have been infected by other cattle.

This is apparent from the government's own data. On 4 July 2013, seven weeks before the start of the first year of pilot badger culls in Gloucestershire and Somerset, Defra held a press conference to launch its strategy for achieving bovine TB free status in England.

All attendees received a 113-page report, *Strategy for Achieving Officially Bovine Tuberculosis Free Status for England*, which laid out Defra's long term strategy for eradicating bovine TB in cattle over the next 25 years through a combination of cattle-based measures and badger culling.

On page 12 of the report was a very interesting chart mapping the evolution of the bovine TB epidemic in Britain from 1956 to 2010.

A series of different strategies were developed from the early 1970s onwards aimed at reducing the badger population both through gassing and trapping and shooting. Up to 1991, Maff had undertaken 734 badger control operations in areas of the country where TB in cattle was thought to have originated from badgers, all but 37 in south west England. Overall, as a result of these culling operations, there was no significant decline in the incidence of bovine TB in cattle or in the proportion of TB infected badgers found in samples examined by Maff.

By the early 1990s the progressive reduction in bovine TB started to stall. Bovine TB incidents began to rise slowly again in south west England, with new outbreaks of TB remaining three times higher than other parts of the country. By 2001 this had taken the level of cattle slaughtered back to 5,000 per year. However even at this level, this was far from a crisis which threatened the very future of the UK livestock industry.

So, we cannot be certain that badgers give TB to cattle, but we do know that badgers and other wildlife have become increasingly infected with TB, as a result of the huge increase in bovine TB in cattle, following restocking after the foot and mouth crisis.

Today there is conspiracy of silence about the huge increase in TB levels following the foot and mouth outbreak in 2001. Understandably, the NFU and policymakers in Defra would rather sweep this sad episode in livestock disease control under the carpet, but the media have also failed to inform the public of the disastrous consequences of restocking cattle without effective TB controls.

Despite the media coverage of the badger culling issue over the last five years, you will be hard pressed to find any reference to bovine TB levels rocketing as a result of short term economic decisions pressed onto the last Labour government by the NFU.

In its 113-page eradication strategy document, Defra dealt with this hugely important episode with a single anodyne line stating:

'**Another consequence of the foot and mouth outbreak was the geographical spread of bovine TB to new areas of England through the restocking of depopulated herds.**'

A consequence of the huge increase in TB in cattle after foot and mouth was the need for NFU and Defra to move the focus away from their catastrophic failures in disease management, which had heaped such misery on the farming industry.

The best way this could be achieved was to find a scapegoat and one came in the shape of the badger. Using its PR machine the NFU went to work organising farmer meeting around the country, lobbying MPs and briefing the media on the growing threat to cattle from the badger. As the disease levels increased, with 40,000 TB cattle slaughtered by 2008, the badger blame game went into overdrive.

The badger has been an innocent victim of political and economic failures in our farming and food production system. Although a minority of badgers have TB, we have no reliable evidence to indicate that they play any significant role in giving it to cattle. What we do know that it takes months for a diseased badger to infect a cow with TB in a confined artificial environment. We also know from field research in England and Ireland that badgers largely avoid cows in both fields and farmyards and cows will rarely graze where there is badger faeces or urine.

The badger might be an easy target for the farming industry and its friends in the government but, to coin a phrase from Prince Charles, to say it is important factor in the spread of bovine TB in cattle is intellectually dishonest.

9

THE BADGER ARMY

'Never doubt that a small group of thoughtful, committed citizens can change the world; indeed, it's the only thing that ever has'.

This famous quote by Margaret Mead, an American scientist and cultural anthropologist, inspired the modern environmental movement that led to the creation of organisations such as Greenpeace and Friends of the Earth.

When the coalition government pushed ahead with the badger cull in 2012, it believed the policy would generate little public opposition beyond a small core of badger protection groups and animal rights activists. However, it failed to take account of the very special place the badger occupies in the conservation and wildlife protection movement in Britain and the impact of the new digital age of communication in empowering people to protest. The advent of smartphones and social media has enabled a new generation to share information and to organise events rapidly.

In April 2013, while campaigning against the cull at Care for the Wild, I was contacted by such a group of individuals who had come together on social media under the banner of 'London against the Badger Cull'. They planned to organise a huge march in London.

Over the next few months, I helped the campaign group link up with key charities including Care for the Wild, League Against Cruel Sports and Network For Animals to support the march and to agree a route with the Metropolitan Police. The event received a major boost when the Queen guitarist and animal welfare campaigner, Brian May, and the actress and founder of the Born Free Foundation, Virginia McKenna, agreed to speak.

Saturday 1 June 2013 was a beautiful early summer day and thousands of people descended on London from all corners of the country to show their opposition to the cull. The march generated considerable media coverage and I was interviewed by the BBC, ITV and Sky News.

Sky News lined me up for a live debate with dairy farmer and NFU spokesman David Barton from his farm in Somerset. David is large imposing man even when you're debating him from a news studio 200 miles away. He made it clear in the interview that by trying to prevent the culling of badgers animal welfare campaigners like me were threatening the very future of his farm. He talked of the impact of bovine TB on the dairy industry and lost no opportunity in pointing the finger of blame at the badger. He claimed that the thousands of people marching through London in protest against the cull policy were largely a combination of anti-farmer and animal rights extremists.

My argument was simple: the cull had no scientific or animal welfare justification and was opposed by a significant majority of the British public, many of whom were sympathetic to the plight of farmers losing cattle from bovine TB.

The march was a great success with thousands of people parading through the streets of London with badger masks and costumes and banners. Brian May and Virginia McKenna made passionate powerful speeches capturing the spirit of the day and the growing public anger over a cruel badger culling policy that was more about politics than science.

Despite David Barton's claims, news footage of the march showed a peaceful carnival family day out with the best of British care and compassion for wildlife on display. But I came away thinking we needed far more than a one-off march in London to generate public awareness and growing political opposition to the badger cull policy. We needed to do something unique with the anti badger cull campaign, we had to create a badger protection movement across the length and breadth of England, with protest marches in every major town and city.

The cull was edging closer. On Monday 27 August 2013 after three years of consultation, delay, and heated political and public debate, the National Farmers Union President announced the start of the pilot badger culls in Gloucestershire and Somerset.

Like many other people in the anti badger cull campaign movement, this moment filled me with both anger and despair. Despite the public opposition, the lack of any credible scientific justification and the cruelty involved, the government and farming industry were hell-bent on killing thousands of badgers.

The start of the culls led to some excitement in the media with many journalists donning their wellies and heading for deepest Gloucestershire and Somerset in the hope of seeing large protests reminiscent of the Newbury bypass campaign and heated stand-offs between badger protectors and the NFU's hired guns, who were on a mission to kill them for £20 per head.

As a leading figure in the anti badger cull campaign, I started to receive calls from the BBC in the late hours of Sunday evening to go head-to-head with the NFU on the badger cull for the headline morning news broadcasts. By midnight interviews had been lined up for Monday morning on BBC 5 live, BBC Breakfast, BBC Radio 4 Today Programme, News 24 and the Vanessa Feltz show on BBC Radio London.

Alone in the early hours of Monday morning, my mind was whirring about how best to use the opportunity to reach millions of people, with a strong clear message on why the badger culls should not be happening. Although many leading naturalists and conservationists had stated their opposition to the cull the BBC had made it clear this was a political controversy and that any BBC broadcaster who spoke out risked breaching BBC guidelines on political neutrality and, ultimately, losing their job. Thankfully, the BBC's leading nature broadcaster, Chris Packham, was not going to be intimidated. In the hours I was preparing to take to the airwaves in defence of the badger, he inspired me by tweeting:

> 'Brutalist thugs, liars and frauds will destroy our wild-life and dishonour our nation's reputation as conservationists and animal lovers.'

Having constantly fought with the government and farming industry over their negligence and deceit in blaming badgers for the spread of bovine TB, these words brought tears to my eyes. They made me more determined than ever to fight this insane, cruel, destructive policy. I responded to Chris thanking him for speaking out and confirming I could use his words when interviewed on BBC news channels in the morning.

For months the NFU press office had been planning its media strategy for the launch of the pilot culls. The culls in Somerset and Gloucestershire were to test whether the method of free shooting badgers at night would be effective and humane. They were not set up to test the scientific effectiveness of the policy in lowering bovine TB and should have run for four years to be followed by review. The NFU press office had lined up senior NFU officers and farmers to brief national and regional journalists. The aim was to saturate the airwaves with the voices of distressed farmers talking of the social and economic impact of bovine TB and the need to kill badgers to reduce the disease.

At Defra, Owen Paterson, and his ministers were also coordinating their media strategy with the NFU by attacking the indecision of the previous Labour government to deal with TB in badgers and telling the public how the coalition was going to take the tough decisions needed to eradicate bovine TB, including tackling the disease in wildlife.

A key element of both the NFU and Defra media strategy was to stoke a climate of fear about the badger. At every opportunity Defra ministers and NFU spokespeople talked of an explosion in badger numbers as a result of legal protection and of the high level of TB in badgers and how they were spreading the disease to cattle in fields and farm yards.

The fact that none of these claims could be backed up by evidence was of no concern; it was a classic propaganda campaign aimed at brainwashing people into believing that anyone opposed to the cull was an extremist who didn't want to protect farmers or deal with diseased badgers.

I arrived at BBC Broadcasting House at 6:15am to take on the government and farming industry, to be met by a young intern, who told me all BBC channels were running the cull as their top news item and that if all went well I would break a new record at the BBC for speaking on a headline news item on all the major BBC National TV and radio news channels within a 45-minute window.

My first interview was on BBC 5 Live responding to the NFU President Peter Kendall, who talked of the desperate plight of farmers with bovine TB in their herds and the need to cull badgers to help reduce the disease. In response to the opening question on why I felt the badger cull should not go ahead, I repeated the 'brutalist thugs and liars' quote from Chris Packham, which I said reflected the huge level of public anger over the pointless slaughter.

Next stop was *BBC Breakfast* where I hit back hard at NFU's claims that badgers had a high level of TB, by stating the fact that none of the animals culled were to be tested for the disease, which took the interviewer by surprise and, hopefully, made people think over their cornflakes. By the time I reached the BBC *Today Programme* studio I was creating quite a wave of public interest on social media and beyond. For the first time across the major BBC news channels, the government and farming industry were being held to account. Evan Davis opened the interview by asking me why the badger had such a special place in the conservation movement compared to other species. I responded by saying it was an animal that suffered greatly at the hands of man as a result of persecution going back hundreds of years and it was now on the frontline in the debate over the future of our countryside and farming.

The anti badger cull movement was not against farmers: we wanted to see a farming industry working hand in hand with nature, rather than destroying wildlife in a drive for higher productivity and profits. Despite the claims of the government and the National Farmers Union, there was no evidence to indicate that killing badgers would lower bovine TB in cattle herds. The interview ended on a humorous note, with Jim Naughtie bringing up the pet badgers Owen Paterson kept as a child (*see Chapter 16, Owen Paterson*). I said they bit him and ran away; clever badgers, they could see what was coming.

On a BBC 5 live debate many farmers called in to say I was trying to destroy their industry by preventing the large-scale culling of badgers. Many called for badger gassing to be reintroduced and for the lifting of the legal protection for badgers. Callers also included farmers and vets who were opposed to killing badgers and pointed out that the spread of bovine TB could far more effectively tackled by focusing on cattle rather than killing wildlife. It was heartening to hear many people call in from the culling zones in Gloucestershire and Somerset to say opposition to killing wildlife was as strong in the countryside as in cities.

My next stop was the Vanessa Feltz breakfast show on BBC London. On this occasion the key point of an NFU regional chairman from Somerset was that we were far too sentimental about badgers and we should accept the need to control their numbers in the same way as rats or other vermin to protect the interests of the livestock and dairy industry. This was a weak argument to make for lifting the protection on a protected species and I tore into his arguments. The response from listeners to the programme was immediate with many clogging the phone lines to lend their support to my case for protecting badgers, in the face of such blind hatred from the farming industry. At one stage Feltz jumped out of her chair and urged me on as I made a passionate defence of the badger.

The end of the interview became quite emotional as I told listeners that I was proud to come from a nation where teachers, vets, nurses, airline pilots were willing to give up their nights to wander dark country lanes to protect badgers in the face of a cruel wildlife destruction policy.

Within a few hours I had reached millions of people with a passionate defence of the badger in the face of well-prepared campaign by the government and National Farmers Union to picture a farming industry in crisis as a result of the spread of bovine TB resulting from the species. However a number of key issues were becoming clear to me in the media coverage of the badger cull issue. Firstly the justifications for the killing of badgers were not being sufficiently scrutinised or questioned by the BBC and other broadcasters.

For example at no point were Owen Paterson, the Defra Secretary of State, or the President of the NFU, Peter Kendall, asked to explain why none of the badgers to be killed were being tested for TB. All the main news broadcasters accepted that Paterson and Kendall would not debate live with me or any other wildlife campaigner on the cull. Throughout the day I found myself arriving at a news studio to learn that Paterson or Kendall were making a swift exit. In some cases, as with Iain Dale on LBC, Paterson was asked directly on air if he would debate with me. He refused.

Having developed something of a reputation in the media as a campaigner against the cull with Care for the Wild (now merged with the Born Free Foundation), I was asked to speak at 'Somerset Against The Badger Cull' protest march in Taunton.

On the day of the protest, 7 September 2013, I took the early train from London only to arrive shortly before the start march in the main park in the centre of the Taunton. I arrived during a downpour and waited patiently in the rain to speak to the hundreds of people crowded into the park. On being invited to the microphone by the chair of the

Somerset Badger Group, Adrian Coward, the sun suddenly broke through the dark sky and the brollies started to come down. I told the story of the incompetence, negligence and deceit at the heart of the badger cull policy and how badgers were being killed for short-term political and economic interests. The reaction from the crowd to my speech was hugely positive and it was soon up on YouTube.

By the time I was making my way home on the train, I was receiving messages on social media asking if I could speak and help organise further marches against the badger cull around the country. My speech in Taunton between the sunshine and the showers, had started the ball rolling on a campaign that would become known as the Badger Army.

Over the course of the next four weeks, I spoke at protest marches in Northampton, Bedford and Kettering. All of these events took place in the evening and people brought lanterns as we marched through the centre of the towns. We generated significant local media coverage and growing public and political interest in the anti badger cull campaign.

By the time the Badger Army reached Brighton on Saturday 2 November over 1,000 people marched along the seafront and through the centre of the town to the Theatre Royal. By now the protest movement was beginning to concern the government and the farming industry.

To capitalise on the growing public anger against the cull, a final Badger Army march for 2013 was held in Bristol on 1 December. Hosted by Care for the Wild and the Somerset Badger Patrol, it attracted over 2,000 people. Leading naturalists and broadcasters Simon King and Bill

Oddie joined me as speakers along with the Bristol Labour MP, Kerry McCarthy.

By now the informal Badger Army had become the largest wildlife protection campaign in Britain and was sparking political interest as well as media headlines. After each march photos and video footage were uploaded to the internet and widely shared on Twitter and Facebook. Diane Bartlett, a member of the Shropshire Badger Group, filmed many of the Badger Army marches across the country and her films which have been viewed by tens of thousands of people in the UK and around the world, have become an invaluable history of the badger protection movement as it's grown in strength and influence.

The Badger Army also effectively tapped into the BBC and independent regional TV and radio networks by holding marches in towns across the country that rarely see protest events. Hundreds of people filling the market square of a town like Dorchester or Ross-on-Wye protesting against the culling of badgers created headline news in the local media and put local MPs views on the issue in the spotlight.

As the Badger Army marched on, from Stratford-upon-Avon to Leeds, and from Exeter to Birmingham, the political pressure on the coalition grew and cracks started to appear between the Tories and the Liberal Democrats on badger culling. By 2014 the badger cull had become a lightning rod issue in Westminster with a Mori poll showing that badger culling was the fifth most common issues of complaint to MPs, above other key economic and social issues, such as child care and taxation, behind only immigration, benefits, housing and the National Health Service. More than half of Conservative and Labour MPs, 54%, identified the badger cull as one of the issues they received most letters on or were approached in their constituency surgeries about.

The pressure was kept up with two debates on the cull in Parliament, on 5 June 2013 and 13 March 2014. A Downing Street petition in June 2013, fronted by Brian May, calling on the government to stop the planned cull, collected 304,244 signatures. It warned that while the cull would kill more than 70% of the badger population in large areas of the country, independent scientific studies had shown that it would do little to reduce bovine TB.

What started to worry backbench Tory MPs and whips in Parliament was the broad coalition of people from all walks of life and backgrounds who joined the Badger Army marches. Doctors, nurses, vets, teachers, architects, retired RAF and commercial pilots – this was not your stereotypical animal rights, far left anti globalisation type movement.

A key leader who emerged in the anti badger cull movement was a perfect example of this new breed of wildlife protection campaigner. Nigel Tolley, a successful restaurateur and owner of a manufacturing business from the West Midlands, threw his weight and energy into organising grassroots support for the anti badger cull campaign. With megaphone in hand Tolley, whose other passions were for sports cars, motor racing and fine dining, would soon be a familiar face to thousands of people in towns and cities across the country, coming down the street leading hundreds of people chanting 'Save our Badgers'.

Many people who protested against the cull, had never seen a badger, except dead by the side of the road. They did not belong to badger protection groups. Many had never protested against anything in their lives before. However they had one thing in common – they were united in their anger over the destruction of wildlife for short term economic and political interests. Many felt powerless when it came to dealing with global issues such as climate change and conflict in the Middle East, but they believed they could make a stand to protect their local wildlife, even if they were never fortunate enough to see it.

10

GREEN MOVEMENT FAILS THE BADGER

While ordinary people across the country have protested against the badger cull, there has been a marked lack of action from the major conservation and wildlife protection NGOs. Greenpeace, Friends of the Earth, WWF, World Animal Protection have done almost nothing to stand up and protect the badger in the face of the cull (and other threats to its existence, such as wildlife criminals and unscrupulous builders and property developers). This is all the more worrying because the fate of the badger is at the centre of a much wider debate about our system of government and the importance of industry and economic growth compared to conservation and the protection of species.

All these issues are of critical importance to the work of Greenpeace, WWF and Friends of the Earth, so why the silence and lack of campaigning and lobbying activity on badger culling? To answer this question we need to look at the changing face of the environmental movement over the last 40 years in Britain and its increasing corporatisation and move away from direct campaigning and activism.

In the 1970s on the streets of London and other towns and cities across Britain, thousands of people were supporting campaigns to protect the environment, animals and wildlife; Trafalgar Square was regularly packed with protesters campaigning against commercial whaling, the Canadian seal cull, vivisection, and nuclear weapons testing.

Pictures of the earth beamed back from space for the first time a few years earlier had showed a beautiful but fragile planet at risk from air and water pollution, dwindling energy resources, radiation and pesticide poisoning. Out of these concerns was born the modern environmental movement. It was a new social movement based on people power rather than wealth and political access, and it borrowed the tactics of the civil rights and anti war and nuclear weapons movements that had become such a strong political force in the United States and Europe.

In 1971, Friends of the Earth was established in Britain. To draw media and political attention to its launch, a handful of volunteers with homemade banners dumped 1,500 non-returnable Schweppes bottles on the front steps of the company's head office in central London. Their message was simple: government and industry should set up a nationwide recycling network. This might not sound very radical today, but in 1971 it was front page news. This type of direct activism had never been seen before and it was a wake-up call for the media and politicians on the growing strength of the green movement.

1971 was a year of great change in Britain and across Europe. The economic boom years of the 1960s had come to an end, students were becoming increasingly radical, the government of Edward Heath was increasingly unpopular and scientific warnings over the damage being done to the planet by mankind were beginning to hit home. Friends of the Earth's founding motto 'Think globally, act locally' really started to resonate in a society which for the first time was willing to take to the streets and fields to change the way government and industry treated the environment and wildlife.

After the media interest generated by the Schweppes protest, Friends of the Earth (FoE) groups sprang up across the country. Over the next few years the organisation grew in strength and influence and soon scored some breakthroughs, including stopping Rio Tinto from digging a copper mine in Snowdonia National Park and banning the import of whale products and tiger skins into Britain.

In the early years of FoE, national groups supported largely autonomous local networks who tackled local problems often by peaceful direct action in their communities. Marches, rallies, boycott campaigns all became tools of the new green people power movement.

As well as the birth of FoE, 1971 was also the year that Greenpeace came to the world's attention. Motivated by a shared vision of a green and peaceful world, a small team of environmental activists set sail from Vancouver Island in Canada to bear witness to the US nuclear weapons testing at Amchitka, a tiny island off the west coast of Alaska. Amchitka was a hugely important refuge for many endangered species including sea otters, bald eagles and peregrine falcons. Although the old boat was intercepted before it reached the weapons testing ground, the campaign generated media interest around the world and Greenpeace was born. The US detonated its nuclear bomb but all further testing ended that year and the island was later declared a bird sanctuary.

Six years later with four members and £800 in the bank, Greenpeace UK was born in a borrowed office in Whitehall. Over the next few years Greenpeace dominated the media with a series of ground-breaking direct action campaigns. It exposed the dumping of radioactive waste by a British ship, *The Gem*, in the North Atlantic; forced the government to abandon a seal cull by bringing its ship *Rainbow Warrior* to the Orkney Islands; and confronted the Norwegian and Icelandic whaling fleets.

The World Wildlife Fund (WWF) was established a decade earlier than FoE and Greenpeace in Geneva 1961 with the mission to stop the degradation of the planet's natural environment and to build a future in which humans lived in harmony with nature. WWF was originally set up to fundraise and provide grants to existing NGOs based on best scientific knowledge and ability to protect endangered species.

As it grew in size and influence, the WWF began to run its own conservation projects and expanded its activities into areas such as sustainable use of natural resources and the reduction of pollution and climate change. Today it is the largest conservation organisation in the world with over five million supporters worldwide and offices in 100 countries. Its annual global income is in excess of £600 million, with 55% coming from individuals, 19% from government sources, and 8% from corporations.

WWF UK was established in 1961 as the first National Organisation in the WWF network. Today it has offices in England, Scotland and Wales and over 300 staff. Its living planet centre in Woking cost over £20 million and was opened by Sir David Attenborough in 2012.

Forty years on the environment movement in the UK has changed out of all recognition from the early direct action campaign days. Today organisations such as Greenpeace, FoE and WWF have millions of members and supporters and generate a combined income in excess of £100 million per year in the UK alone.

Gone are the days of borrowed offices and small volunteer teams. Today these organisations employ thousands of staff in large modern offices. No one can doubt their invaluable work and achievements in lobbying government and industry to create a more sustainable green based economy, but much of their ability to energise and mobilise people to campaign directly for habitat and wildlife protection has been lost along the way.

Today, these organisations more resemble the corporations they were created to challenge than environment protection campaign groups. They employ hundreds of administrative staff, with teams of managers in marketing, legal HR and PR departments.

As their supporters and corporate donor base has increased, generating more income, they all employ large numbers of staff in fundraising roles to keep the money coming through the door. Street-level volunteer activists have all but disappeared from these organisations, in favour of professional employed street chuggers or call centres, whose only focus is to recruit new supporters and generate more income. The supporters, though numerous, have little or no influence over the priorities of the NGOs they are funding. All these matters are left to faceless boards of directors, chief executives and presidents and their senior management teams.

As these organisations have become larger and more corporate in their structure and outlook they have adopted a philosophy of compromise. As their leaders come under constant pressure to pay salaries and campaign bills and maintain corporate sponsorship agreements or access to governments, any bold demands and risky campaigns are shelved. Despite these compromises aimed at increasing income and influence the most serious problems these organisations are aiming to eliminate, from climate change to deforestation and habitat and wildlife destruction, are getting far worse not better.

Large NGOs have been increasingly restricted in their campaigning activities by changes to criminal justice, anti-terror laws and charity regulation, which governments have been only too happy to use as a way of stifling legitimate public protest and opposition to their environment policies. And corporations and trade bodies have increasingly turned to PR and legal firms to influence public opinion or threaten any attempt by NGOs to blacken their reputations by calling into question their influence or access to the highest levels of government.

In the early 1970s WWF, Greenpeace and Friends of the Earth would have been on the forefront of the anti badger cull campaign in Britain. They would be mobilising their activists to take to the streets and the fields and would be leading the debate on the disastrous cruel badger cull in the media, Parliament and the courts. However in the new green corporate world they inhabit, they now remain largely silent on the issue.

Greenpeace might have led the campaign to ban seal culling in Shetland in the 1970s but now it is too nervous to take on the government and the farming lobby on badger culling.

Friends of the Earth was willing to publicly criticise Owen Paterson for his failure to address the threat of climate change during his time as Environment Secretary, but it gave him a free pass when it came to the mass killing of badgers.

WWF talks boldly about its campaigns to protect endangered wildlife in Africa and other far flung parts of the world, but remains strangely silent when it comes to objecting to a policy which could see the local extinction of a protected species which has lived in our isles for half a million years.

To all these organisations the badger cull issue has become too toxic and politically sensitive. They have no appetite for taking on the lies or propaganda of the farming lobby or senior officials in Defra and their political masters.

To embark on such a policy is now seen as too risky for the brands, corporate sponsors or relations across key government departments. They know the badger cull is wrong on scientific, humaneness and financial grounds and is just the type of issue which brought them into existence in the first place, but they have compromised themselves into a corner and there is no escape.

One large animal welfare and wildlife protection charity has been far more vocal in their opposition to badger culling than the major environment NGOs, but this has proved a hazardous path.

The RSPCA is Britain's most recognised and respected animal welfare charity with over 1500 staff and an income of over £100 million a year. The RSPCA prides itself in being the oldest and largest animal welfare organisation in the world. The charity states that its mission is by all lawful means to prevent cruelty, promote kindness and alleviate the suffering of animals.

In January 2012 Gavin Grant was appointed chief executive of the RSPCA. He had worked at the RSPCA as director of campaigns between 1988 to 1991, before moving to the

Body Shop as its corporate communications manager and becoming chairman of the PR company Burson-Marsteller.

Grant decided he wanted the RSPCA to campaign more on important wildlife protection and animal welfare issues such as badger culling, fox hunting, horseracing, and live animal exports. Taking a high profile position on these issues was a high risk strategy for the RSPCA, but Grant was confident it would be rewarded by greater public support and donations.

Although the RSPCA had been heavily involved in tackling badger persecution through its wildlife crime inspectors, until Grant's arrival it had taken a low profile on the badger cull controversy. During his time as chief executive between January 2012 and February 2014, Grant put the RSPCA at the forefront of the campaign against the cull. In his view it was the RSPCA's duty to lead opposition to a cruel policy that mistreated a protected species.

Soon Grant was leading public meetings and protests against the badger cull across the country. He stated that farmers backing the cull should be named and shamed and voiced his support for voluntary boycotts of milk from badger cull farms. He also supported an RSPCA 'vaccinate not exterminate' campaign in the *Metro* newspaper in London, which led to 119 complaints to the Advertising Standards Authority. The ASA ruled that the use of the word exterminate was inappropriate and misleading and prevented the RSPCA from running the advert in other publications.

Under its new chairman Sir William Shawcross, the Charity Commission, which regulates charities, soon had the RSPCA in its sights and launched an investigation into its political campaigning activities on the badger cull. The RSPCA defended its actions as according with charity law but the pressure started to mount on Grant and the board of the RSPCA to pull back from such a public show of opposition to the culling of badgers.

Unfortunately for Grant it was not only his fierce campaigning on badger culling which got him in trouble with the Charity Commission, but also his views on live animal export and fox hunting. He even started to take on the powerful racing industry over the number of horses injured and killed at major events like the Grand National.

To many animal welfare campaigners Grant was a breath of fresh air. For once a chief executive of a leading animal protection charity was willing to shake up the political and media establishment and take on their powerful friends in the farming and hunting lobby. Grant's problem was that he was not heading a small animal rights charity but a national institution, whose patron was the Queen. Soon a debate raged in the media, with accusations that the RSPCA had been taken over by animal rights extremists and money that should be protecting cats and dogs was being wasted on political campaigns. In 2013 the Archbishop of Canterbury, Justin Welby, confirmed that he would break a long tradition and decline the vice presidency of the RSPCA, which was seen as a snub to a charity mired in political controversy.

The situation was made worse when a memo from the RSPCA's deputy chairman, Paul Draycott, was leaked to the media. The six-page memo to his fellow trustees painted a picture of a charity which had become embroiled in controversial political campaigns without considering the consequences for its reputation and income. Draycott painted a picture of an organisation in crisis with falling income, an exodus of talented staff, and increasingly nervous sponsors and business partners.

The leak was seized upon by organisations like the Countryside Alliance and NFU, who called for new leadership and an end to the RSPCA's anti badger cull and fox hunting campaigns. The RSPCA's chairman, Mike Tomlinson, told the media that Grant had his full support, but the writing was on the wall. By February 2014 less than two years after he was appointed Gavin Grant stepped down as chief executive. Within months most of the senior management team he brought into the RSPCA to lead his new style of animal protection campaigning had also gone.

Grant was controversial and his style of management and leadership was not to everyone's liking, but he was bold and courageous when it came to taking on powerful vested interests to protect animals. Since his departure the RSPCA has struggled to define its role as an animal welfare and protection organisation. It has now been audited twice by the Charity Commission, on its prosecutions and its governance, and has only recently appointed a new chief executive.

Despite maintaining its opposition to badger culling, the issue is far lower down the organisation's list of priorities. The RSPCA has not been completely silenced, but it is no longer a leading campaigning organisation, at least when it comes to standing up for badgers.

The badger cull debate has shown both the strengths and weaknesses of the modern environment and wildlife protection movement. In many ways the public anger caused by the cruel indiscriminate culling of badgers for short-term political and economic interests should have united major environment and animal protection NGOs in Britain in condemnation.

At a time when many have lost touch with their supporters and become too risk averse in order to protect their reputations, funding and government access, the badger cull was a perfect opportunity to get back to peaceful direct action. Instead, the battle to save the badger is being left to small charities with tiny budgets and in many cases a handful of staff, like the Badger Trust.

If the legacy of the corporatisation the environment movement is the loss of species like the badgers from large areas of the English countryside it will be unforgivable.

11

DEFENDED BY AMATEURS

As the Secretary of State, Owen Paterson, pushed forward with the badger cull in the summer of 2013, the National Farmers Union started to engender a climate of fear within the farming industry over the possibility of extreme animal rights activists sabotaging the cull by threatening and intimidating the farmers and cull contractors carrying out the policy. Paterson was happy to play along with this game, even though there was little evidence to support these claims.

A small number of the more extreme direct action elements of the animal rights movement had been active on social media making threats against farmer and cull contractors, but they did not reflect the vast majority of the peaceful law-abiding badger protection movement. However the actions of a few did give the NFU the excuse to apply for – and obtain – a legal injunction under the Protection

From Harassment Act 1997 aimed at protecting farmers, occupiers of land and those participating in the pilot culls from unlawful acts. Many aspects of the injunction were completely legitimate in terms of trying to prevent protesters from harassing or intimidating those participating in the cull or unlawful entry or damage to property. However it was also a blatant attempt by the NFU with the co-operation and support of the government to prevent any peaceful protest or monitoring of the cull contractors in the cull zones.

Under the terms of the injunction no group of individuals could gather anywhere in the vicinity of homes or businesses involved in the culls, including on public footpaths. They were also prevented from using torches, cameras, video equipment anywhere near the cull zones. As the exact locations of the cull zones were not released, this effectively meant that large parts of west Gloucestershire and west Somerset would be protest-free zones throughout the periods of the culls.

With the support of other key wildlife protection charities opposing the cull, the Badger Trust challenged the wide-ranging nature of the injunction and its impact on the right to peaceful protest at the High Court. In a ruling in the High Court on 22 August 2013, Mr Justice Turner accepted the right of the NFU to prevent intimidation and harassment of farmers involved in the badger cull, but he also made a number of important changes to the injunction to maintain the right for peaceful protest in the cull zones.

The badger cull policy was now becoming an important civil rights issue and although the NFU was entitled to use the law to prevent threats and intimidation against its members, this could not come at the price of preventing law-abiding citizens from going into the fields to peacefully express their opposition to the policy. The High Court ruling came as a huge relief to many in the badger protection community and paved the way for a new ground-breaking wildlife protection movement to be established to bear witness to the cull.

In the months running up to its start, ordinary people passionate about their local wildlife formed groups in Gloucestershire and Somerset to organise patrols in the cull zones.

With valuable professional training support from Jordi Casamitjana from the International Fund for Animal Welfare and funding from wildlife protection charities, including the RSPCA, League Against Cruel Sports and the Badger Trust, the Badger Protection Movement was born.

A husband and wife, Jeanne and Nick Berry, helped establish Gloucestershire Against Badger Shooting to campaign against the badger cull and to organise wounded badger patrols. In Somerset, the local badger protection group formed Somerset Badger Patrol to organise peaceful opposition in the fields. Adrian Coward and Vanessa Mason, two key members of the local badger group, helped get patrols up and running together with a former BBC natural history film maker, Amanda Barrett.

Soon both groups had recruited hundreds of supporters and had developed protocols and training programmes to support the largest wildlife protection patrol movement ever seen in Britain.

As public anger grew against the badger culls, people from all walks of life and backgrounds volunteered to join the movement and patrol the culling zones in Somerset and Gloucestershire. Teachers, vets, bankers, nurses, former airline and military pilots all came together to make a stand to protect wildlife.

When the culls began at the end of August in 2013 many newspapers and news channels despatched their reporters to Gloucestershire and Somerset in search of stories that would show violent conflict between badger protectors, farmers and cull contractors reminiscent of the anti road building protests of the 1990s. To their surprise, they found themselves joining a professionally organised highly committed peaceful wildlife protection movement that avoided conflict but was determined to bear witness to the cruel killing of badgers.

When BBC *Inside Out West* joined a Wounded Badger Patrol in the Somerset cull zone on Monday 30 September 2013, it found teachers, firemen, vets, librarians, computer programmers, doctors, surveyors and opera singers all joining the movement. And they came from all over the UK from Wigan, Swansea, Brighton, Cornwall, Manchester, Oxford and London. Caroline Allen, a vet and national spokesperson on animal issues for the Green Party, told the BBC:

'We want to be here to show solidarity with the protesters. We know that the cull is completely unscientific, unethical and will be inhumane. As a vet, I'm completely against the cull. I'm disgusted that some of my professional organisations have shown their support for badger culling.'

Will Ricks, a surveyor from Ross-on-Wye, told the BBC he met a huge mixture of people on the patrols:

'There are teenagers up to 70 year olds on the patrols. In a field at midnight I bumped into two people who had driven down from Manchester for the evening. They stayed until 2am and drove back to work the next day. Were all people who have a strong view that the cull is completely the wrong thing to do and it's completely pointless.'

Soon the media across Britain and around the world were carrying reports and images of people in high visibility jackets carrying torches, using night vision googles and thermal imaging cameras as they walked public footpaths in the middle of the night, to monitor the activity of badger cull contractors.

Reacting to the fears planted by the National Farmers Union over threats and intimidation to their members participating in the cull, the police brought a huge number of officers and equipment into the culling zones in 2013.

The NFU placed officials in the police control rooms and played a hands-on role. A police officer filmed in Gloucestershire in 2013 told an anti badger cull protester that his details would be passed on to the NFU, which the officer said might chose to pursue a private prosecution and this video was shown on *The Guardian* website

The police and NFU then admitted that representatives of the cull company HNV Associates were in the police control room along with an NFU representative. The police defended this decision by saying it enabled real time information about the location of the cull contractors; the NFU was not directing the police operation

The Somerset and Gloucestershire Police Commissioners raised concerns on this issue and the matter was also raised in Parliament. By 2014 the procedures were changed and NFU and cull contractors' representatives were not given automatic access to the police control rooms for the culls.

Hundreds of officers were taken from key frontline duties; to monitor the Wounded Badger Patrols.

Four-wheel vehicles, specialist radio and night vision equipment and even helicopters were despatched to the badger cull zones. The cost of this policing operation soon started to run into millions of pounds as helicopters hovered over teachers, nurses and members of the Women's Institute at cost of over £1,000 per hour.

The Gloucestershire and Somerset police promised throughout the cull to deal with both protesters and badger cullers fairly, not showing favouritism to either side. Policing the culls at night was undoubtedly a difficult task for the two constabularies involved, but soon concerns were soon being raised that the police were acting as a security force for the farmers and cull contractors.

When protesters were stopped and interviewed by police officers they often found their details were being passed on to the National Farmers Union, effectively making the police a security service for a farming trade body. Police officers in Gloucestershire even started to hand out NFU produced leaflets warning protesters of the dangers of infringing their civil injunction. The leaflets stated that a breach of the civil injunction may lead to summary arrest' without a warrant and 'you are hereby put on notice of the injunction.'

The NFU even became involved in the sacking of a civil servant from Defra's Rural Payments Agency who criticised the badger cull policy and the NFU President Peter Kendall on social media. The NFU lodged a complaint with Defra who in turn mounted an investigation which led to her dismissal on gross misconduct grounds. Defra refused to release any correspondence with the NFU on the issue.

On the ground in the cull zones, peaceful protesters in the badger patrols sometimes found themselves confronted by contractors and farm employees involved in the cull, who were aggressive and intimidating. In a number of incidents patrol group members were pushed and swore at and their vehicles were damaged or forced from the road.

According to a media report, a woman in her 60s was given a police caution for assault after tussling with a cull protester in Somerset in October 2013 *The Guardian* reported that the victim, Sharon Doyle, said she and a friend were checking a map at the side of the road after hearing reports of shooters in the area. She was subjected to verbal and physical abuse and her car keys were wrestled from her, leaving her bruised and in a state of shock. *The Daily Telegraph* reported the arrest of a 38-year-old man in the Gloucestershire cull zone in September 2013 for attacking an anti cull protester, Meg Joes, and for damage to her vehicle, which had both its wing mirrors broken off.

Many of the cull contractors were little more than poorly trained pest controllers, who, with minimum supervision, were left to wander the countryside at night in search of badgers to kill, despite the potential public safety risks. In some cases, cull contractors followed badgers onto golf courses and public footpaths and took shots near areas of residential housing or where members of the public were out walking their dogs. The contractors were under strict operating rules to cease any shooting should they come into close contact with protesters or members of the public, but these rules were not always followed.

On one occasion, a cull contractor left his open vehicle to chase a protester with his loaded gun unsecured on the back seat. A protester, Madeline Buckler, took possession of the weapon to show how the contractors were taking risks with public safety. She was arrested, but the case never came to Court as the cull contractor withdrew his statement.

In March 2014 David McIntosh, once known for appearing in the TV series *Gladiators* pleaded guilty to crashing a van full of dead badgers into a bus stop in Gloucestershire city centre at 1am on 29 September 2013. He admitted to driving without due care and attention and not having a valid driving licence. He was employed as part of the cull contractors to deliver dead badgers for incineration.

A number of protesters from Gloucestershire against Badger Shooting protested outside the court hearing. They were angered by evidence submitted during the hearing which showed the police were in direct contact with McIntosh by radio at the time of the accident and that he was able to get a job delivering dead badgers for incineration, without any checks being made of his driving licence.

In August 2014 a cull monitor working for the Animal Health and Veterinary Laboratories Agency broke his silence in an article in the *Sunday Times*, painting a picture of poorly trained trigger-happy cull contractors compromising public safety and falsifying records. He sent his concerns in a report to Defra and had a follow-up meeting with officials, but no action was taken.

Alongside the wounded badger patrol movement, another group which played a major role in opposing the badger cull movement in the fields was the Hunt Saboteurs Association (HSA). The HSA was formed in 1964 to take direct action against hunts in the countryside rather than lobby for parliamentary reforms to end blood sports. By 1965 it had groups established across the country including in Devon, Somerset, Birmingham, Hampshire and Surrey.

Over the decades many in the hunting community and their friends in the media and Parliament have attempted to paint a picture of the HSA as a violent anarchist movement, which threatens the safety of huntsmen their horses and hounds. However, this could not be further from the truth. The HSA brings together people from all walks of life who believe they must make a stand to prevent the cruel destruction of wildlife for pleasure. In the small number of incidents where HSA members are arrested, it's usually for breach of the peace or aggravated trespass as a result of entering a farmer's field, or blowing a hunting horn or spaying lemon oil in a wood. The key aim of the HSA is to observe hunts and intervene to tip the scales in favour of the wild animal that are being hunted and persecuted.

The HSA is a professionally run organisation that trains its volunteers to avoid conflict with hunters and any harm to their animals. It researches and keeps meticulous records on the hunts it monitors and disrupts. Over the decades, despite often facing intimidation and violence, the HSA has brought to public attention the brutal reality of hunting, and its activities have influenced public and political opinion on fox and stag hunting and hare coursing.

The HSA has taken a more direct action role than the badger patrol movement in trying to disrupt the badger culls. It has been willing to push the boundaries of the NFU injunction on trespass and access to the cull zones and has played a key role in protecting badger setts and locating and removing cages used to trap and shoot badgers. It has made great use of night vision and thermal imaging equipment and established base camps in the cull zones, where it has built an effective intelligence network on the location of badger setts and movement of cull contractors.

Despite the best efforts of the *Daily Mail* to portray the HSA as dangerous radicals seeking to infiltrate the peaceful badger protection movement, I have huge respect for its role in peacefully opposing the badger cull in the fields. Whenever I have spoken to groups of HSA activists, I have been impressed by their dedication and professionalism and willingness to put themselves in the frontline of wildlife protection. They have also played a significant role in supporting the badger protest movement in towns and cities across the country helping to maintain media interest and political pressure.

A more controversial but important figure in the badger protection movement is Jay Tiernan, who runs the 'Stop the Cull' campaign. The NFU listed Tiernan in its injunction aimed at preventing farmers and cull contractors being threatened and intimidated for killing badgers.

Tiernan has a long history of being involved in animal rights campaigns, some of which have landed him in trouble with the law for trespass and threatening behaviour, particularly the campaign in the 1990s against animal testing by Huntingdon Life Sciences. He soon became a bogeyman in the media for his past brushes with the law and his willingness to use direct action against farmers and cull contractors to bring an end to badger killing.

Tiernan has always been highly suspicious of me because of my government and corporate lobbying background, particularly my time working in the plant science industry. He was also very aggrieved by comments I made outside the High Court in August 2013, when I supported the NFU's right to have an injunction protecting its members from direct threats and intimidation. Despite our differences and my concern over some of his campaign methods, I admire his tenacity and ability and his willingness to battle with the NFU in the courts and the media to keep the cull in the public eye.

In January 2015, Tiernan was found guilty in the High Court of nine of breaches of the NFU injunction including filming a farmer shooting and killing a badger and trespassing on private land. He was given a six-month prison sentence suspended for two years and ordered to pay

£25,000 in costs. If the NFU hoped this would result in him disappearing from the anti cull movement they were soon disappointed.

In June 2015, Tiernan targeted Caffè Nero over its sourcing of milk from the badger cull farms for its coffee shops. In response to what Caffè Nero called serious and credible threats to its team member and premises, it decided to source its milk from outside of the cull zones. This angered farmers and soon became an issue of debate in Parliament. The media jumped on the story and soon Tiernan was debating the issue of milk sourcing and badger culling on the *Jeremy Vine Show* and with the NFU Vice President on *Newsnight.*

Although I disagreed in public with Tiernan's tactics of using social media to threaten direct action protest against Caffè Nero and other food retailers who were sourcing badger cull milk, I admired the way he used the media interest to put the cull back in the headlines. His campaign against Caffè Nero also helped generate a much needed focus on the role of the food service and food retail sectors in the badger cull debate.

The actions of the badger patrol groups, hunt saboteurs, and direct action protesters such as Jay Tiernan, could not save every badger, but there is no doubt they saved hundreds if not thousands of badgers from being cruelly shot, and helped make the cull more complex and costly to implement.

As the culls have continued and expanded the police have improved their methods and have pulled away from the confrontational heavy-handed tactics which caused much anger and concern for protesters in 2013. They have learned that the vast majority of people who choose to enter the fields at night to show their opposition to the killing of badgers are caring, compassionate and law abiding. Despite the fears generated by the NFU and government ministers, very few arrests have been made as a result of the actions of protesters. Indeed, most of the threats and intimidation has been from those involved in the culling operation not those opposed to it.

I can think of no other country in the world, where people would react to a government policy to kill wild animals by organising patrols and walking footpaths in the all weather conditions until the early hours.

The commitment and courage of this movement is best summed up by the example of Sue Chamberlain. In May 2013 a meeting against the badger cull took place at Dorchester town hall organised by the RSPCA and Brian May's Save Me charity. Gavin Grant of the RSPCA and May spoke at the packed meeting in advance of the first year of the pilot badger culls. Chamberlain was in the audience and the meeting changed her life. Motivated by anger over the planned killing of badgers she went on to play a key part in setting up Dorset for Badger and Bovine Welfare.

Over the next two years she organised numerous meetings, set up a social media campaigns and helped recruit and train new volunteers to map badger setts and carry out badger protection patrols. She laid down the foundation of a badger protection movement which was ready to spring into action when the culls were extended into Dorset in August 2015. From the first night of the culls Chamberlain – Sue to me – was in the front line, organising patrols and speaking to the media. Anyone who met her was impressed by her drive and commitment, this was all the more remarkable because Chamberlain was fighting a private battle against cancer at the very time she was fighting to protect badgers. In January 2015 she sadly lost her battle against the disease, but she has a left a legacy for protecting badgers in Dorset which will live on in her memory. To me she sums up everything that is special about the wildlife protection movement in Britain today: her care, compassion and courage characterise a new generation of wildlife protectors.

12
BBC BIAS

As the badger cull progressed the BBC became heavily criticised by the badger protection movement. In my view some of this was unfair, particularly as its regional networks had given extensive coverage of the badger protection patrols in the culling zones and the anti badger cull protest marches in towns and cities. However in some cases the BBC did fall well short of its commitment to maintain impartiality as a public service broadcaster.

On May 31 2013 the BBC placed an article on its website by David Gregory-Kumar entitled 'How did the Irish badger cull play out'. Its focus was a link between the mass culling of badgers in Ireland and a drop in the level of bovine TB in cattle. The report carried the lines:

'When I talk to local farmers about bovine TB, so many of them point out that culling badgers in the Republic of Ireland has helped control the disease. The data does seem to back that up with the numbers of

infected cattle falling in Ireland and slowly rising in England. Long term, an affordable vaccine is the way forward. But, the lesson from the Republic of Ireland is that a badger cull, along with other measures, can help control the disease until then.'

This belief that culling badgers in Ireland has lowered bovine TB in cattle has been regularly used by the government to justify culling in England. Addressing the Oxford Farming Conference in January 2014, for instance, Owen Paterson said: 'Since the Republic of Ireland has been culling, they have seen a reduction in the disease of some 20%.'

However as we have no scientific proof that badgers can spread TB to cattle, the BBC was completely wrong to state that badger culling can help control the spread of disease, despite Paterson's political spin.

Tom Langton, a respected environmental consultant and badger protection campaigner, made a formal complaint to the BBC. His key concern was that the misleading BBC report was being used by Defra ministers and NFU staff to make inaccurate claims about the impact of killing badgers on lowering TB rates in cattle. Over six months the complaint was batted about the BBC, during which time the corporation continually twisted and turned and tried to do all it could to make Langton give up. Initially, indeed, the BBC editorial unit rejected the complaint, but Langton appealed to the BBC Trust, which oversees the broadcaster. Eventually, after the BBC Trust intervened, the BBC Editorial Standards Committee partially upheld the complaint under its code on accuracy. In March 2014, the

BBC Editorial Standards Committee stated that while TB in Irish cattle was at historically low levels this could have been caused by many factors. It ruled that while the article had not knowingly misled the public:

> 'The language used in the article had not been sufficiently precise as it suggested that the badger cull is a factor in helping control bovine TB when it is not possible to be definitive on this issue.
>
> 'While the data did show a decline in the number of cattle infected with TB in Ireland, there was no conclusive evidence to show that the badger cull has been categorically responsible for any of this decline and so it was inaccurate to say that, along with other measures, it can help control the spread of the disease.'

Put simply, the BBC concluded (rightly) that there was no proof that badgers spread bovine TB to cattle.

Many in the badger protection movement hoped that this acceptance of a major editorial failure would ensure the BBC reported the cull more evenly in the future. But within weeks the corporation had caused even more anger by giving Princess Anne an audience of eight million on BBC One's *Countryfile* to mount a defence of badger gassing as a way of controlling bovine TB.

In early April 2014, *The Sunday Times* alerted me to an interview Princess Anne had given to Tom Heap for *Countryfile* at her Gatcombe Park estate in Gloucestershire. The BBC billed the interview as a 'frank and wide ranging look at issues affecting the British Countryside' and took the unusual step of releasing an excerpt to other broad-casters a few days before it was broadcast on *Countryfile* on Sunday evening.

On travelling home on a Friday afternoon through Euston station, I looked up at the big video screen to see a Sky News report with Princess Anne suggesting that gassing was a much nicer way of killing badgers. Suddenly my phone started ringing for media interviews. By the time the programme was broadcast on Sunday evening, it had already caused a large amount of controversy and hundreds of people had complained to the BBC about Princess Anne's views on badger gassing and the programme's failure to provide the Badger Trust with a right of reply.

The BBC responded by saying that Princess Anne had run her Gatcombe Park estate for almost 40 years and had lost 15 of her White Park cattle to TB in the previous two years and therefore had strong views on the need to cull badgers to reduce bovine TB. However, the corporation failed to explain why out of a wide ranging 15 minute inter-view it released only her comments about badger gassing to other networks in advance of *Countryfile's* broadcast.

The BBC has a special arrangement for royal interviews where it can share excerpts with other broadcasters before the interview goes out. The interview was over 15 minutes long so why take the few minutes she spoke about gassing badgers to share with other broadcasters and nothing else? Princess Anne also spoke about eating horses which is equally controversial – if not more so – but this segment was not shared prior to broadcast. Blaming the badger for spreading bovine TB has become a mantra among the political, media, farming and veterinary establishment in Britain with the strong support of senior members of the Royal Family, particularly Prince Charles and Princess Anne.

In June 2014, the BBC published an independent report on the BBC's coverage of rural areas across TV, radio and online. This was a significant opportunity to take a wide-ranging look at how the BBC dealt with controversial rural issues such as badger culling and fracking. The report was written by Heather Hancock, a former land agent and chief executive of the National Parks Authority, who had chaired the BBC's Rural Affairs Committee.

Titled *BBC Trust Impartiality Review: BBC Coverage of Rural Areas in UK*, it was broadly sympathetic to the farming community. It stated: "Overall, audiences and stakeholders thought the BBC did a good job on a difficult and complex on-going story." But it noted that stakeholders felt that too much attention was paid to cull protesters and not enough to farmers coping with the social and economic impact of losing their cattle to bovine TB. Hancock wrote: 'People asked where were the pictures of sick badgers with TB, or infected cows being shot, or a distraught farming family coming to terms with the loss of their animals.' She felt that the BBC's coverage had focussed too much on cull protesters and images of healthy badgers and not enough on explaining the scientific and agricultural context of the cull. The report stated:

'Even on the BBC website, the story was introduced by photographs and a drawing of a badger. The content analysis bore out this finding. Badgers were by far the most dominant visual motif in coverage, accounting for more than 50% of coded visuals in BBC and non BBC coverage. By contrast, cows and cattle had far less prominence. The wider point about undue focus on conflict in the countryside was also borne out by the 25% presence of images of protesters.'

Making a general recommendation to the BBC to improve its coverage, Hancock warned: 'The emotional impact of pieces was perceived to lead to unintended bias in how the audience perceives a story.' Publication of the report in June 2014 led to media reports about BBC rural affairs coverage having too many fluffy badgers and being too much in favour of the anti cull lobby. Although this could not be further from the truth, the report did influence the future direction of rural programming at the BBC, which finally led to the commissioning of *Land of Hope and Glory.*

Land of Hope and Glory was a three-part series broadcast on BBC 2 in March 2016 following a year in the life of *Country Life* magazine. It was produced by Jane Treays, who had made the fly-on-the-wall documentary *Inside Claridges.* The series was billed as an opportunity for viewers to meet heroic country house owners and defenders of the landscape, architectural heritage, and field sports, and a key figure in its production was Mark Hedges, *Country Life*'s editor. He was given a significant amount of editorial control over the programme, in return for giving the BBC inside access to his publication for 12 months. Viewing figures hit an impressive 1.8 million, a publicity coup for *Country Life*, whose sales jumped by 35%.

But the problem with allowing Hedges this level of editorial control over a BBC programme became clear following the broadcast of the first episode of the series, which included a major focus on Maurice Durbin, a Somerset farmer with bovine TB in his herd. From inaccurate statements about the level of TB in badgers to misleading remarks about the number of badgers and their need for protection, the programme went out of its way to demonise badgers.

On number of occasions a clearly distressed and emotional Durbin appeared to be encouraged to make angry remarks about badgers and those who seek to protect them. In one scene the camera zoomed in on Durbin after one of his cattle tested positive for TB for him to say 'bloody badgers'. In another, he talked of kicking badgers and of 'do-gooders telling farmers what to do' and how he was placing himself at personal risk by talking of the need to kill badgers.

Mark Hedges increased the tension by talking of farmers committing suicide as a result of bovine TB and how the population of badgers had increased thousands of times since they were given protection in the 1970s, which in his view they no longer deserved.

After the programme was broadcast, I wrote on behalf of the Badger Trust to the BBC's Director General, Tony Hall, complaining that *Land of Hope and Glory* was encouraging farmers and landowners to take the law into their own hands and illegally kill badgers. The Badger Trust's chairman, Peter Martin, complained that the BBC had abandoned any pretence of balance on the bovine TB issue and was guilty of 'institutional bias.'

Hedges responded: "There are always two sides to a story, and we are proud that we have enabled the farmer's story to be told at last. A single-issue group should not be allowed to bully this BBC for doing that."

The Times published a full page spread on the row by its environment editor Ben Webster under the headline: 'Badger lovers set about the BBC'. On LBC, host Nick Ferrari and I heatedly debated Hedges' claim that I was a 'badger bully' trying to silence the BBC for telling the truth about the impact of bovine TB on farmers.

The BBC dismissed claims of institutional bias and encouraging wildlife crime, stating that *Land of Hope and Glory* was not a current affairs programme, but an observational documentary about people who live and work in the countryside.

Following its broadcast, however, Hedges acknowledged that he had overruled a suggestion by Treays to include an interview with Brian May to provide some balance, remarking: "I said it would be pretty boring because he would say what he thinks and I would say what I think but what has never been shown is what happens to the dairy farmer."

The debate over the BBC and badgers will go on and I don't underestimate the difficulties the corporation faces in dealing with a complex and highly charged issue such as bovine TB and badgers, which so clearly divides public, political and scientific opinion.

However it's clear that the government and its supporters in the farming and landowning community continue to have a significant influence over the BBC's editorial and rural affairs coverage. In my view this is leading to significant failures when it comes to ensuring the controversial issue of badger culling is treated with the proper accuracy and impartiality as required by the BBC's Royal Charter editorial guidelines.

The BBC should investigate and challenge public opinion, but its most important obligation to licence payers is to present the truth.

13

ILLEGAL CULLS

Today in Britain badgers are regularly killed by farmers, landowners, hunt masters, game keepers, property developers, badger baiters and a variety of sick individuals, who gain some perverted pleasure from subjecting wildlife to pain and suffering. However, the problem goes far wider than illegal criminal activity by such individuals. It goes right to the heart of government and the farming industry, which all too often turn a blind eye to illegal badger persecution.

In 2014 the Welsh government held a major summit on bovine TB in Cardiff, bringing together academics, policymakers and representatives from the farming and veterinary industry in the UK and around the world. Defra's Chief Scientific Advisor, Ian Boyd, was asked for his greatest regret about bovine TB policy in the past 30 years. His answer? The Protection of Badgers Act. Boyd argued that legal protection had allowed numbers of badgers to rocket, allowing them to become a major reservoir for bovine TB infection in cattle.

The fact that the Chief Scientific Advisor to Defra was willing to go on the record at an international conference with such a statement shows how ingrained this view had become within the tight cadre of senior policymakers in Defra. If they believe the badger does not deserve legal protection, what hope is there that their political masters will enforce the law on badger persecution?

In early October 2013 I came face to face with this brutal reality at the end of the first six weeks of the pilot badger culls in Gloucestershire and Somerset. Asked by Sky News to give a short interview on public opposition to the cull at Secret World, a wildlife rescue centre in Somerset, I spent an afternoon filming with reporter Isabelle Webster, against the backdrop of some rescued badger cubs.

Awaiting my train at Taunton later, I received a call from Webster, who told me that after completing an interview, a farmer in Somerset, who mistakenly thought the camera was switched off, had bragged about gassing badgers on his farm – and added that neighbouring farmers were doing the same.

Was this important? Yes, I said. Gassing is not an approved killing method as it is considered inhumane. It is also illegal. No badgers can be killed without a licence from Natural England: the farmers were acting illegally. The interview showed what many in the badger protection movement feared: that farmers were organising themselves into groups to carry out the large-scale, illegal killing of badgers on their farms, both inside and outside the official culling zones. It shed new light on Defra's just-announced statement that there had been a significant reduction in the estimates of badgers in cull zones.

Owen Paterson's department had just announced a three-week extension to the amount of time marksmen would have to kill badgers in Somerset, because they had proved harder to find. In west Somerset, the population, which had been estimated at 2,400, was put at 1,450. Instead of there being 3,400 badgers in Gloucestershire, the number was estimated at 2,350. In an infamous TV interview, Paterson denied that the cull was incompetent because the authorities had failed to estimate correctly the number of badgers, or to kill 70% of them, and now had to extend the cull to do so. Wasn't this, a BBC interviewer asked, 'moving the goalposts?' In what became a much-mocked retort, Paterson responded:

'The badgers moved the goalposts. We're dealing with a wild animal, subject to the vagaries of the weather and disease and breeding patterns.'

Disease, breeding patterns and the weather were the causes of the decline, according to the Secretary Of State. Yet here was a hint that something else may have been to blame as well, or instead. Sky News decided to back to Somerset the following morning to set up a number of interviews. After a busy day filming, Webster called me the following evening to say she had recorded further interviews with farmers and now had evidence of illegal gassing using a hosepipe and vehicle engines on 14 farms in Gloucestershire and Somerset. This would go into an exclusive news investigation, with the identity of the farmers being protected for legal reasons. She asked me to be interviewed for the piece alongside Owen Paterson.

Deciding to be bold, I put the blame for illegal gassing firmly on the government and the NFU. I also raised fears that the government was not only failing to ensure the perpetrators of illegal badger gassing were brought before the courts, but was also looking to adopt gassing as a future method for badger removal.

In his interview, Paterson admitted that illegal killing undermined the licensed culling operation as it could lead to disruption of badger colonies and increased risk of pertur-bation. He could not bring himself to condemn the action of the farmers as illegal. Instead, he said 'they were unfor-tunate' and that random unlicensed culls could lead to an increase in the spread of bovine TB, which needed to be avoided. I felt that his statement gave the impression that he sympathised with farmers having to take the law into their own hands when it came to badger control on their land. The NFU refused to put up a spokesperson for the inter-view, but in a written statement it denounced any illegal badger killing.

Sky News broadcast the investigation on Wednesday 9 October, generating significant interest. The following morning, I was invited to join Nicky Campbell in the BBC 5 Live studio at Broadcasting House in London to debate the badger cull with Jan Rowe, a representative of one of the badger culling companies. Rowe joined the interview from his home in the West Country, which was probably for the best in view of how heated our exchange became. The interview began predictably, with me condemning badger culling on scientific and humaneness grounds and Rowe defending the need for the policy on the basis that it would lower bovine TB rates in cattle.

Rowe remained calm in his responses until Campbell brought up the issue of illegal badger gassing and asked me who I thought was responsible. I responded by saying that it was an open secret in the farming community that illegal badger gassing was taking place across the West Country and that Rowe probably knew some of the farmers involved. I went on to say that Rowe was aware of organisations like the Badger Welfare Association which were openly convening farm meetings across the West Country to discuss badger gassing techniques and that it was no good being in a state of denial on the level of illegal badger persecution activity by farmers and landowners.

Rowe went into a rage and denied any knowledge of illegal badger gassing or which farmers might be responsible. He flatly rejected my comments.

By this time, I had decided to throw caution to the wind and use every opportunity I could to put the spotlight on Owen Paterson and the NFU for not doing enough to stop farmers taking the law into their own hands.

By chance, I was scheduled to make a speech on Saturday 11 October at a badger protest march in Witney, David Cameron's constituency, together with some interviews with local media. It was an opportunity to take the case for firm action to stop illegal badger killing to the Prime Minister's backyard. On the Saturday around 500 people turned in Witney up for the badger march, with banners, badger costumes and even a set of goal-posts, in tribute to Owen Paterson's comments about the badgers moving them.

The march wound its way up Witney High Street past David Cameron's constituency office and onto Witney Green. In glorious Autumn sunshine, I leapt onto a park bench, megaphone in hand, and stressed why the badger cull was a disastrous failure on scientific, cost, and humaneness grounds, and how it was only taking place because of a pre-election deal between Cameron and the NFU to shore up Tory support in rural constituencies.

However I saved the best for last by coming back to the Sky News investigation on the illegal gassing of badgers. Prior to attending the march in Witney, I had written an opinion article for Politics.co.uk titled 'The conspiracy of illegal badger gassing,' and this became the focus of the closing part of my speech.

I told the hundreds of people on Witney Green that the media had lots of fun following the statement by Owen Paterson that the 'badgers had moved the goal posts', but that beyond the humorous front pages and cartoons and spoof video games, the announcement by Natural England of a 66% decrease in badger numbers in the cull zones over the previous 12 months was a matter of very serious concern that merited further investigation. I questioned the Natural England's assertion that the massive decline over such a short period of time was due solely to a cold winter, shortage of food, and disease. No expert in the field of badger behaviour and population studies was willing to accept that these factors alone could explain such a huge drop in numbers.

Although illegal persecution of badgers had taken place by farmers and landowners for many decades, the TV news investigation had proved that farmers were now organising themselves into groups to carry out widespread coordinated illegal gassing of badgers using carbon monoxide from vehicles. They were also using social media to share best practice and sett locations and even forming professional networks to hold meetings in rural communities to openly discuss the killing of badgers.

I laid the blame for this huge increase in wildlife crime at the door of Owen Paterson and the leadership of the NFU. They were, I said, allowing farmers to act with impunity when it came to killing badgers. I also raised fears that the government was pushing ahead with researching gassing as an option for culling badgers, giving an effective green light to criminal activities.

After social media whipped up media and public interest in the illegal gassing of badgers, the backlash arrived. It came in the form of a letter from solicitors acting on behalf of the NFU who recorded all the statements I had made in the broadcast and print media on the issue of illegal gassing of badgers in the previous four days. I was not certain how much NFU was paying its solicitors, but it must have been considerable because someone in the law firm had diligently recorded every word I had uttered on TV, radio in the press and even on Twitter. The solicitors claimed that my media statements had unfairly tarnished the reputation of the NFU and its President Peter Kendall and that some of my comments could be considered libellous in a court of law.

On receipt of the letter, I contacted the solicitors and told them it was not my intention to hold Peter Kendall directly responsible for the gassing of badgers, although I remained of the view that he and the NFU needed to do far more to condemn this activity by farmers and landowners.

Following further consultation with their clients, the solicitors contacted me again to say they would like to recommend that I issue a public apology to the NFU President, to set the record straight that I did hold him personally responsible for illegal badger gassing. I agreed to do this. I was then asked to issue a similar statement to Owen Paterson.

I was very surprised to see a firm of solicitors representing the NFU making such a request on behalf of a cabinet minister and I refused. I told the solicitors if Paterson wanted to take up the issue with me, he could do so directly through government legal channels, not through the NFU.

When making the apology to Peter Kendall, I made it clear that I remained of the view that illegal badger gassing was a widespread problem and more needed to be done to tackle this criminal activity. I also asked that when publishing my apology to Peter Kendall, the NFU issue a statement condemning the illegal gassing of badgers. When the apology was published, the NFU press office made sure it was widely picked up by the farming press, but unsurprisingly failed to issue any statement condemning illegal badger gassing to accompany my apology. I never heard from Owen Paterson.

A few days after the apology was published, I was contracted by a reliable source within Defra to say that at high level meeting the Chief Scientist, Ian Boyd, and the Chief Vet, Nigel Gibbens, had considered the 66% decrease in badger numbers in the cull zones, estimated by Natural England. My source told me no official record was kept of the meeting, but both Boyd and Gibbons accepted that illegal killing of badgers by farmer and landowners in the cull zones both before and during culling could have been a factor in the plummeting number of badgers.

On receiving this information, I took to Twitter and asked Boyd and Gibbens if such a meeting had taken place and if they considered illegal badger killing to be a factor for the decline. Over the next 24 hours I heard nothing, but then within 20 minutes of each other Boyd and Gibbens responded with an identical statement on Twitter stating that no such discussions had taken place and they did not consider illegal killing of badgers to be a factor for the significant decline in badger numbers in the cull zones.

The Secretary of State was not willing to state publicly that farmers and landowners were illegally killing large number of badgers. I suspect to this day that Paterson's use of the phrase 'the badgers moved the goalposts' was not a silly utterance but a clever political move to distract the media from more closely scrutinising the key factors behind the drop in estimated badger numbers in the cull zones. If they had done so, they would have soon realised that illegal killing had to be a major factor for such a drastic fall in such a short period of time.

The badger was rightly given legal protection over 40 years ago in the face of widespread cruelty and killing by baiters and diggers, but all these years later it has many more enemies both in the countryside and in Whitehall and Westminster.

This is a political scandal and a wildlife protection disaster. If we continue to see the farming industry, Defra policymakers and their political leaders turn a blind eye to badger persecution as the official cull continues, we can no longer be certain of the size of the badger population. There is already significant doubt about is extent.

In 2005 the Joint Nature Conservation Council estimated the total UK population of badgers to be around 288,000. A Defra-commissioned study in 2011 put the population at around 400,000, but this was based on counting entry and exit points of badger setts and extrapolating those across large parts of the country.

As many as 50,000 badgers are thought to die on the roads every year. During their anti badger persecution campaign week in 2015 police forces across the country confirmed that as many as 10,000 badgers a year are killed through illegal persecution. Many badger cubs die during a dry hot spring when they cannot easily get to their main food source, earthworms. Outbreaks of disease such as the gut ailments *Isospora melis* and the highly pathogenic *Eimeria melis* can also devastate numbers. (Dry spring weather and gut parasites may seem to be relatively innocuous but together they can lead to severe malnutrition and widespread mortality). At the time of publication (July 2016), 4,000 badgers have been killed in badger culling and if culling is extended to 20 or more areas over the next four years this could reach 100,000 by 2020.

As we have such poor information on badger numbers we could see local extinction of the species from areas of Britain.

In the 21st century, we should not be losing more large mammals. In the past 2,000 years Britain has lost the lynx (400 AD), the brown bear (1,000 AD), the beaver (1526 AD) and the grey wolf (1680 AD). The otter was wiped out in many counties for decades but, thankfully, returned after a concerted campaign. Our history suggests there is no room for complacency on the badger.

14

TB BURGERS

TB is an extremely unpleasant illness in humans, causing a phlegmy cough, night sweats, fever, and fatigue, and weight loss. Defra likes to remind us of this, and warns that *M.bovis* is not a disease of the past and a risk of infection remains through consumption of dairy products that have not been pasteurized (there has been a gourmet trend for unpasteurised milk and cheese) and through occupational exposure to TB animals and carcasses. There are 30 to 40 cases of *M.bovis* in humans in the UK every year, according to Public Health England.

Whenever the media reports the number of TB cattle slaughtered, they often state the animals have been destroyed, leaving the public with the impression the carcasses are incinerated to prevent them entering the food chain. But this is not so.

In May 2013, I was contacted by a highly effective but secretive anti cull campaigner on social media known as Spartacus. His information, obtained using the Freedom of Information law with the government's Animal Health Veterinary Laboratories Agency lit up a murky side trade in TB meat in the UK.

Every cow that fails the tuberculin skin test (called a reactor) is slaughtered to stop the spread of bovine TB . Defra uses market prices to compensate the farmer for the value of the cow based on age and sex, pedigree status and type (beef or dairy). All slaughtered cattle are subject to a post mortem examination to check for TB type lesions.

Up to this point no surprises, but what Spartacus had to tell was mind-boggling: up to 20,000 TB cattle were entering the food chain every year and Defra was profiting from the business.

At first I found this hard to believe, but close inspection of the data showed Spartacus was absolutely right. At a time when the Defra Secretary of State, Owen Paterson, was playing up the need to cull badgers to protect human health, his own department had a sideline selling meat from TB-infected cows without any traceability or labelling to food service companies.

This was how the system worked (and, at the time of writing, still works): when a TB reactor cow is slaughtered and compensation is paid to the farmer it becomes the property of the state. If an inspection of the carcass in the slaughter house reveals TB lesions in more than one organ, it is declared unfit for human consumption and destroyed. But if the TB reactor slaughtered has TB lesions in the lymph nodes or just one organ, the area is removed and the rest of the cow is considered safe to enter the food chain. In 2012

Defra figures confirmed that 88% of the cattle slaughtered fell into this 'safe to eat' category and over 20,000 entered the food chain generating around £12 million for the UK treasury, which was set against the £40 million paid out to farmers under the TB cattle slaughter compensation scheme.

To make matters worse despite the trebling of the number of TB cattle entering the food chain between 2002 and 2012, the government had not acted on the advice of its own Food Standards Agency to reduce the risk of the public consuming TB meat. Concerned by the rising number of TB cows entering the food chain, the FSA's Advisory Committee on Microbiological Safety of Food recommended full labelling and traceability of all TB meat entering the food chain and its pre-heating to kill off any TB bacteria (in lesions too small to be seen by the human eye)

For over a decade Defra has refused to act on these recommendations.

The risk of catching TB from eating raw or undercooked meat from a TB-infected animal is very small. Nonetheless, I was of the view that the food industry had to be very careful. Between 1994 and 2011 there were 570 reported human cases of *Mycobacterium bovis* infection (about 33 a year), according to the Health Protection Agency. Most of these were in people aged 45 or over. These cases could be down to consuming unpasteurised milk or dairy products, contact with TB infected cattle or consumption of TB meat.

To avoid anyone catching TB countries such as the United States and Germany have clear labelling rules and cooking instructions for all meat from TB infected cattle.

Switzerland and Italy do not allow any TB infected meat to enter the food chain.

Having worked at the Ministry of Agriculture through the BSE crisis and dealt with the aftermath of the dioxins and the Sudan 1 food dye incidents in the food industry, I was all too aware of the dangers of causing an unnecessary food scare with the TB meat story. However, I felt the public had a right to know that their money was being spent buying TB infected cattle, which were then being sold back to them in food products, without any labelling. I also believed it was important to show that while Owen Paterson was calling for the widespread killing of badgers to protect human health, his own department was selling TB meat without even following public health recommendations from the Food Standards Agency.

Before taking the story to the media, I worked with Spartacus to seek some answers from the major food retailers and pet food manufacturers on the use of TB meat in their products. The responses we received left us in no doubt that we should publicise the story. Tesco and Sainsbury confirmed that they would not sell TB meat as it would not be acceptable to their customers on food safety grounds. Whiskas, a leading pet food manufacturer, confirmed it would not use TB meat in its cat food products, because it would not be acceptable to pet owners.

Compiling all the information we had collated on the TB meat trade, I approached Channel 4 News, who were interested in the story. But as if often the case with news broadcasters today, they were not willing to spend the time putting together the key facts and seeking the necessary responses from Defra and the Food Standards Agency.

Disappointed but not deterred, I contacted Jonathan Leake, science editor of *The Sunday Times*, who took a keen interest in what we had found and started to research an in-depth piece. I was hoping we might get a small piece in the paper, which would allow us to build media interest in the story the following week. However luck was on my side as unbeknownst to me Owen Paterson had just arrived in New York, in advance of a speech he was going to make promoting 'Food is GREAT Britain' at the Fancy Food Show in New York.

When Leake took the story to his editor, Martin Ivens, the decision was quickly taken to put the story on the front page. *The Sunday Times* is put to bed only a few hours before publication and I received notification from Leake at 11pm on Saturday that the story would be on the front page the following day, 30 June 2013. His advice was to wake up early and stay by the phone, as he expected the story to lead the news the following morning.

At 6am I rolled out of bed and put on the TV to find Sky News and the BBC running TB meat at the top of their bulletins. By 7am I had been inundated by interview requests from broadcast and print media. My first in-depth interview was with Channel 4 News, which had suddenly woke up to the importance of the story.

By the end of the day the TB meat story was all over the media in the UK and around the world and I no doubt put an end to any plans the Secretary of State might have had for some sightseeing in New York.

By Monday morning *The Mirror* was running a front page story 'TB Burgers' focusing on the sale of TB meat in schools. By this time Defra had gone into major damage control mode. It released a statement saying:

'This is irresponsible scaremongering. The Food Standards Agency has confirmed there are no known cases where someone has contracted TB from eating meat. All meat from cattle slaughtered due to bovine TB must undergo rigorous food safety checks before the meat is passed as fit for consumption. As a result, the risk is extremely low, regardless of whether or how the meat is cooked.'

Throughout my interviews, I did not seek to cause panic, but I did point out that killing badgers was not benefiting public health and that if the government truly wanted to reduce the risk of *M.bovis* increasing in the human population, it should close down its TB meat sideline. My view was that if TB meat was not good enough for a tin of cat food, it should not end up in meat products fed to school children, hospital patients or our armed forces.

Owen Paterson felt betrayed by *The Sunday Times* over the TB meat story. Only a few months before he had been courted by Rupert Murdoch over a private dinner with Nigel Farage in London to discuss a future EU referendum. On his return from New York, Paterson wrote to the *Sunday Times* editor Martin Ivens, criticising the paper for working with an anti badger cull campaigner on a story which, he said, was intended to whip up a food scare. He detailed Defra's long term TB eradication strategy and the importance of culling badgers for this to succeed. He ended by calling on *The Sunday Times* to cease from writing any further stories raising food safety fears over the sale of TB meat. Thankfully Ivens was having none of it and Leake wrote a follow up story the following week looking at the sale of TB meat to France.

A final twist to the story came in March 2014, when Owen Paterson was fighting a rearguard action in the cabinet to extend the badger cull in the face of stiff opposition from Nick Clegg and the Liberal Democrats. Using data from Public Health England, newspaper stories started to appear raising fears over the increasing danger of humans catching TB from their pets. The focus was on two people in Newbury who had developed TB from cats in 2013, following an outbreak involving nine cats in west Berkshire.

Although the report on the incident stated: 'There is no evidence to suggest that *M.bovis* from cat to human is anything other than a rare event,' a *Daily Mail* front page warned: 'Four cat owners catch TB from their pet.' The article included a handy guide on the chain of TB infection from cattle to humans. In the middle was an image of a savage looking badger which was put in the frame as the most likely source of TB infection for the cats, who had in turn infected their owners. A report in *The Daily Telegraph* on 30 March 2014 stated:

> 'The TB outbreak among domestic cats, which has now spread to humans, reinforces the need for a badger cull to help tackle the disease, the government has suggested.

'Officials at the Department for Food and Rural Affairs (Defra) said bovine TB continued to cause untold misery to beef and dairy farmers and the recent cluster of cases among pet cats in the Home Counties, demonstrated why it was necessary to address the issue head on.'

It is most likely that the cats contracted the disease by eating infected rodents.

The timing of this TB health story as Owen Paterson was fighting to extend the badger culls could be just be a coincidence, but somehow I doubted it. My suspicion was that Defra felt it had to damn the badger if the struggling cull policy was to command any significant support. Yet support was ebbing away.

15

OWEN PATERSON

Troubled may be the best adjective to describe Owen Paterson's two years at Defra. From the very start, he was suspicious of the ministry, suspecting that its policymakers had been captured by wrong-headed green groups.

In some quarters there was scepticism that Paterson was the right politician to run the environment department. He was a Euro-sceptic hard-right member of the English squirearchy. He was aligned to the Countryside Alliance, the climate change sceptic network of Nigel Lawson and his controversial Global Warming Policy Foundation, and the journalist and science writer Viscount Matt Ridley, who happened to be his brother in law.

Within months of arriving at Defra in September 2012, he had caused controversy by refusing to meet his Chief Scientific Advisor, Ian Boyd – and for claiming at the Conservative Party conference that there could be advantages to climate change.

The Conservatives' most prominent environmentalist, Zac Goldsmith, was quoted as saying:

'The appointment is odd, if the Conservatives wished to retain their green credentials then it would have been better to appoint someone who didn't dismiss environmentalism as a left-wing issue.'

However, Paterson represented a rural constituency and had long taken an interest in countryside issues, and was a strong supporter of farmers and other agricultural businesses.

After reading history at Corpus Christi in Cambridge, and with the benefit of fluency in French and German, Paterson had worked his way up to become managing director of the British Leather Company in 1993. In 1997, he became MP for North Shropshire. As shadow agriculture minister under Michael Howard, Paterson campaigned for the dairy and fishing industries, and took an increasing interest in bovine TB and badgers, even travelling to the United States to discuss TB management in deer and cattle.

In the 2010 coalition, he became Secretary of State for Northern Ireland, where he pushed for the desegregation of schools between Protestants and Catholics. In Ulster, he maintained his interest in farming, working closely with the NFU and Countryside Alliance in his rural seat.

On entering Defra in 2012 he felt he had to take on the powerful green lobby, which he called the 'Green Blob'. What better way to demonstrate that than by defeating the greens who wanted to protect cuddly badgers – while at the same halting the catastrophic spread of bovine TB?

Although Paterson had to postpone the cull in 2012, he remained very optimistic about the project's chances of success. In May 2013, before the cull had even begun, he suggested that it should be expanded to 40 new areas of the country. 'We want to reduce the incidence of disease [bovine TB] to less than 0.2 per cent of herds a year,' Paterson told *Sunday Times* science editor, Jonathan Leake, in an interview. 'It will take 20-25 years of hard culling to get to that.' He also talked about lifting the legal protection on the species to make it easier for farmers to kill badgers on their land as a pest control measure, without a Natural England licensing and monitoring system.

Indeed during his tenure at Defra, Paterson maintained his considerable enthusiasm for the cull, which was supported by Downing Street. His enthusiasm was not dimmed by any setbacks. He was not put off by the condemnation of animal disease experts who stated unequivocally that the cull would make little difference to controlling bovine TB, and could even make it worse. Nor by the widespread public protests which greeted the start of the cull in August 2013. Nor by the problems in correctly estimating populations of badgers to ensure enough of them had been shot. Nor, apparently, was he troubled by the notion of farmers, at some point in the future, killing 'pest' badgers on their own land.

Despite his enthusiasm, problems with the cull continued to mount. In February 2014, the BBC's science correspondent, Pallab Ghosh, revealed that Defra's Independent Expert Panel - which Paterson had set up – had found that the free shooting of badgers during the pilot culls in

Gloucestershire and Somerset in 2013 was not effective. Nor was the free shooting method, the cheaper method favoured by the government, particularly humane: up to 18% of culled badgers took longer than five minutes to die.

In 2014, with Badger Army protests continuing up and down the country, Paterson's fellow coalition MPs started to become concerned about the rising public anger towards the cull. With his party's poll ratings in free-fall, the Lib Dem leader Nick Clegg for one was no longer willing to be associated with such an unpopular policy. In spring 2014, despite having a number of MPs in the south west, where farmers were pushing for the cull, Clegg made it clear that the Lib Dems would not support its extension into Devon and Cornwall, at least until the flaws in its effectiveness and humaneness highlighted by the independent expert panel had been resolved.

Following other perceived failures such as Defra's response to horsemeat being found in burgers and lasagne, Owen Paterson was becoming increasingly isolated. The NFU and Countryside Alliance were quick to come to his aid, making it clear to David Cameron that Paterson was the authentic voice of the landowning farming and hunting community, which remained of critical importance to the Tories electoral fortunes in rural seats. But it was not enough to expand the cull, as he had hoped.

On 3 April 2014, Paterson announced to the Commons that culls would continue in Gloucestershire and Somerset for their four-year term, but there would be no extension of the programme into new counties, at least not yet. He re-stated his strong belief that the badger cull, combined with 'stringent' cattle-based measures, would reduce bovine TB, telling the Commons: 'We are clear that culling needs to be part of the answer as there is no other satisfactory solution available at the moment.'

He thanked the Independent Expert Panel for its findings, promised to incorporate them into the culls, and scrapped the panel. Saying there had been lessons to learn from the first year of the cull, he announced he was changing policy on badger vaccinations. While the coalition had cut the number of vaccination trials from six co two, Paterson decided that it was now worth injecting the vaccine into badgers on the edges of TB hotspots – to create a buffer zone of immunity that would stop the disease from spreading. Inside the cull zones, the contractors would wipe out 70% of the badgers and in the surrounding areas the badgers would be vaccinated (there was no mention of there being a reservoir of TB in other wildlife, such as deer and foxes).

Despite Paterson's confidence, the political problems with the cull intensified. By summer 2014 the Badger Army was in full swing and local Badger Trust groups had organised more than 20 protest marches in towns and cities across England, including in the Prime Minister's constituency. The Badger Trust was granted a judicial review of Paterson's decision to continue the pilot culls without the supervision of an independent body. The judicial review failed, but it brought the lack of effective monitoring to the public's attention and further increased the pressure on the British Veterinary Association to oppose free shooting on humaneness grounds.

Finally, when David Cameron reshuffled his cabinet on 15 July, Owen Paterson was sacked as Environment Secretary. According to reports in *The Daily Mail* and *The Daily Telegraph*, Paterson did not go quietly, demanding a face to face meeting with the Prime Minister rather than the customary exchange of letters. In a stormy discussion in Cameron's Westminster office which could be heard in the corridor outside, Paterson told the Prime Minister that sacking him was a smack in the teeth for the country's 12 million rural dwellers – and demanded: 'Do you actually want to win the election?'. Quoting a 'well-placed source,' the *Daily Mail* reported Paterson as saying: 'I have not been afraid to take on the greens on everything, from fracking to GM foods, the badger cull, even bees!... Sack me and all the green groups... will be celebrating.'

In an article for the *Sunday Telegraph* on 20 July, Paterson said that while the Prime Minister was free to choose his ministers, he (Paterson) had 'misgivings' about the enduring power of what he called 'the green blob':

'By this I mean the mutually supportive network of environmental pressure groups, renewable energy companies and some public officials who keep each other well supplied with lavish funds, scare stories and green tape.'

He singled out the 'anti capitalist agitprop groups' Greenpeace and Friends of the Earth, who he claimed were only interested in building up their income and influence by myth and scaremongering on issues such as pesticide and bees, fracking and GM crops. He neglected to mention that one of the issues that worried the Prime Minister was the public anger generated by Paterson's handling of the badger cull and the growing impact of the Badger Army. The badger cull had been a two-year headache for him at Defra, bringing thousands of people on to the streets and causing political rifts all the way up to the Deputy Prime Minister.

In a Sky News debate during the 2015 election campaign Cameron remarked of the badger cull: 'It's probably the most unpopular policy I'm responsible for. I know it is very unpopular, culling badgers.' He added: 'But I profoundly believe that part of the way of trying to create some parts of the country that are TB-free is to do this. It's very, very difficult but I believe it is the right thing to do.'

John Randall, the Deputy Chief Whip, told me before the reshuffle that badger culling had made Paterson's position very difficult. Cameron had appointed him to show decisive leadership on the badger cull and to reduce its political impact. On both counts he had failed.

By now, Tory MPs across the country were bombarding the Prime Minister with concerns about the way the policy was being mishandled and its possible impact on the 2015 election. However, this was not the result of a highly-funded professional campaign by Greenpeace, Friends of the Earth or WWF; in fact none these influential environmental groups took any role in the anti badger cull campaign. Instead it was run on a shoestring budget by a small organisation (the Badger Trust) and a committed group of activists.

Making maximum use of both conventional and social media, it has become the largest wildlife protection street movement in Europe. More than 300,000 people signed a Downing Street petition, and there have been over 35 marches in towns and cites across England which brought thousands of people on to the streets, in the largest rolling protest movement seen in UK for decades. There has been a huge amount of media coverage, a High Court challenge, and three debates in Parliament.

The Badger Trust puts the protection of the environment and wildlife above economic growth and profits. All its money comes from individual supporters, Badger Trust groups, donations, legacies and training courses. We receive no grants from government or funding from FTSE100 companies. Not all organisations enjoy such freedom.

16

VETS' DILEMMA

Owen Paterson's successor was Liz Truss, an education minister. In March 2015, after months of delay and a personal intervention from the Prime Minister, she finally agreed to meet me and other senior representatives of the Badger Trust: Peter Martin, Dave Williams, Pat and Jeff Hayden.

Over the course of my career, I have met with a number of Defra Secretaries of State and ministers and I was familiar with the routine of being ushered from the waiting room at Nobel House into the Minister's office, to be granted the standard 45-minute audience. Most politicians who reach the cabinet have a certain confidence and air of authority about them; you might not agree with their views, but you can see why they have been successful in politics.

What struck me about my first encounter with Liz Truss is that she displayed none of these traits. She is one of the least confident and charismatic politicians I have ever met.

Truss grew up in a left-wing family in Glasgow and Leeds. She graduated from Oxford University before working as an economist. She entered Parliament in 2010 and became a junior education minister in 2012. Unlike a number of her predecessors at Defra she had no close links to the farming or food industry and, seemingly, little interest in environmental issues. In person she appeared to be a perfect example of a modern career politician: ambitious, focused and very presentable, but – it seemed to me – with no passion, gravitas or depth of character.

What I found difficult to comprehend when we met was her poor understanding of the complexities of the bovine TB and badger issue. She showed a complete lack of understanding or any real interest in the issues we put on the table. Her responses to the concerns put by Peter Martin, chairman of the Badger Trust, and I appeared to be the briefing lines provided by her officials. As expected, she talked of the rising cost of bovine TB to the taxpayer and the social and economic impact of the disease on farmers. The example of Ireland was used to justify the need to kill badgers in England and she emphasised that the reservoir of disease in badgers needed to be tackled.

She made no attempt to explain why none of the badgers killed in the culls was being tested for TB and became visibly nervous when we talked of the £5,000 per dead badger cost to the taxpayer.

The private secretary sat loyally next to her keeping note of the issues raised, but her policy advisors sat at the far end of the table, leaving a visible chasm between themselves and their political master.

One key figure Truss kept bringing up during the meeting was Defra's Chief Vet, Nigel Gibbens. On a number of occasions she mentioned that the Chief Vet was fully behind the badger cull; this was clearly being used as a shield to deflect any criticism that the policy had no scientific or animal welfare justification.

Yet Nigel Gibbens did not attend the meeting in person, even though it was scheduled weeks in advance. Perhaps he decided to stay away on the basis that he could well have new political masters following the general election the following month, who would reverse the badger cull policy for all the reasons we put to Truss. Perhaps he had other business.

The only bright spot for us was that Truss agreed to make a public statement on the need to tackle badger persecution. Although no such public statement was made, she did write to me immediately after the general election on 7 May 2015 confirming that badger persecution was a national wildlife crime priority for the government and that those responsible should be brought before the courts.

The meeting was a largely fruitless exercise, but it did show the influence of the Chief Vet and the wider veterinary profession over the direction of the badger cull policy. When it comes to opposing the killing of badgers the politicians and farming lobby are easy targets, but the motives of the veterinary industry are often overlooked.

Most people would consider protecting animals and promoting animal welfare as being the primary aim of the veterinary profession. Our contact with the profession is usually as a result of our companion animals or from childhood memories of reading books by James Herriot or from

TV programmes like *Vet School* or *The Supervet*.

The reality of the veterinary profession is far more complex than dedicated professionals simply working to heal and protect animals. Veterinary surgeons have many roles, some of which are far more controversial, such as working in animal research, greyhound and horse racing, or circuses. Vets also play a key role in all aspects of the modern intensive livestock industry, from the treatment of animals on the farm to disease control and their slaughter for food.

Across the profession vets face ethical dilemmas, but they are often more vexing in the livestock industry. Changes in animal production and farming practices and the drive to meet the ever growing demand for food are constantly putting pressure on animal welfare standards and increasing the threat to wildlife.

Livestock vets are largely dependent on their farm clients for an income and as such few are willing to speak out against badger culling. The British Cattle Veterinary Association (BVCA) is a powerful force within the wider British Veterinary Association (BVA) and it has ensured the voice of livestock vets controls the BVA's position on the badger culling.

Science and innovation play a key role in the veterinary industry, but this does not seem to apply to badger culling. Despite the many uncertainties over the spread of TB from badgers to cattle and the growing scientific evidence that killing badgers is both inhumane and does not significantly lower bovine TB in cattle, the veterinary industry continues to provides crucial support for the continuation of the policy. It took the BVA almost two years and much painful deliberation in its Ethics and Welfare Group to reach the conclusion that the free shooting of badgers is inhumane and should cease. However, under huge pressure from its livestock members, the BVA simply bowed to pressure from the government and the NFU; who pushed aside its concerns that an end to the use of this cruel killing method would significantly increase the cost of future badger culls.

A number of vets have even become vocal campaigners for badger killing. Den Leonard, a livestock vet from Cheshire, is a prime example. In September 2014 Leonard and another Cheshire-based dairy farmer, Phil Latham joined me for a panel debate on the badger cull at the Labour Party conference in Manchester, chaired by Dawn Primarolo, Deputy Speaker of the House of Commons.

Leonard is a hard-working vet who passionately believes that badgers are a key culprit for the spread of bovine TB in cattle. Leonard rarely smiles and comes across in public as a very serious figure continually on the verge of an explosion of anger when it comes to badgers. Nevertheless he remains a strong voice among livestock vets and is a perfect example of why the debate on badger culling has not evolved within the profession. Ripping into a largely hostile audience in Manchester, he put the blame for the spread of bovine TB in cattle on the badger protection movement, which, he said, had prevented politicians from taking the necessary action to kill large number of badgers in TB hotspots. His views were echoed by Phil Latham, who claimed that TB was seeping into cattle in Cheshire as a direct result of the movement of TB-infected badgers into the county.

Another vet championing badger culling is Roger Blowey, a retired livestock vet who also worked for the Central Veterinary Laboratories. Now called the Veterinary Laboratories Agency, this executive agency of Defra carries out animal disease surveillance, diagnostic services, and veterinary scientific research for government and businesses.

In February 2015, Blowey and some farming colleagues, including a number who were directly involved in the culling operation, wrote to an academic journal for vets, *Veterinary Record*, reporting a significant fall in the number of cattle testing positive for TB inside the Gloucestershire culling zone. The authors claimed the badger cull might be the reason for the reduction. Blowey made a number of media appearances and other public proclamations to support this claim. Despite being based on anecdotal, uncontrolled and incomplete data, it gained some traction among cull supporters. The NFU president Meurig Raymond used it in his speech to the NFU's annual conference in February 2015. Liz Truss and one of her deputies, George Eustice, the farm minister, also quoted Blowey's evidence at various opportunities, while stressing that it was anecdotal.

Blowey finally had the wind taken out of his sails during a panel debate at the Badger Trust's annual general meeting in April 2015, when he and John Blackwell, President of the British Veterinary Association, argued strongly for the badger cull. At the event at the University of the West of England, Blowey took to the stage with graphs claiming to show clear trends in falling cattle TB rates in the badger cull zones. In the middle of his presentation, Professor John Bourne – also on the panel – leapt on stage and stopped his presentation. Bourne then turned to the audience to announce that Blowey had once been a pupil of his and stated that despite his respect for him, he was talking complete nonsense. He then tore into Blowey's case for culling as only a teacher to a pupil could. Bourne pointed out the danger of making assumptions based on anecdotal evidence and Blowey's willingness to jump to conclusions to support his point of view.

Although also a vet, Bourne was scathing about the veterinary industry's continued support for badger culling in the face of mounting evidence that it was ineffective, inhumane and, last but not least, very expensive.

Bourne blamed incompetence and negligence within government for allowing bovine TB to become so widespread. From an over-reliance on the TB skin test, to restocking cattle after foot and mouth without TB testing, and a reluctance to slaughter whole herds when TB broke out, he painted a picture of constant failures in TB reduction policies going back 40 years.

Bourne added that of the 11,000 badgers killed in the randomised cull, very few had late-stage TB with visible lesions who could spread TB via their urine – despite many vets claiming the number of badgers in this category could be as high as 40%. Bourne made it clear that TB could spread very quickly in livestock and that 94% of bovine TB infections were due to cattle to cattle infection, not badgers.

Another voice of reason on the panel that day was Mark Jones, a leading figure in the Born Free Foundation. Both a vet and a respected wildlife protection campaigner, Jones is strong voice within the veterinary profession against badger culling. He told the audience that vets had a vital role to play in reducing bovine TB from providing clinical services and advice to farmers to guiding the direction of government policy in the areas of cattle testing, biosecurity measures and risk based cattle trading. He said the public expected the veterinary profession to uphold the highest standards of welfare for wild as well as domestic animals and this was of key importance when it came to the killing of badgers.

Jones challenged John Blackwell, President of the BVA, to remove all support from any future culling operations that used free shooting. Looking uncomfortable, Blackwell refused to make the commitment. He said the BVA had reached a decision that free shooting was inhumane and should cease, but would not withdraw support for the badger culling policy on this basis.

At this point I challenged him that too many vets were financially dependent on their farm clients to take a balanced view on the animal welfare and ethics concerning badger culling. Blackwell strongly denied this.

In January 2016, Jones, Bourne and other leading animal health experts sent a briefing paper on bovine TB to all the members of the BVA Council. It laid out the key scientific and animal welfare issues around bovine TB reduction and called on the BVA to review its policies on TB control and to improve information and advice to vets and farmers. The authors called on the BVA to reassess its support for the badger cull policy on both scientific and humaneness grounds. It remains to be seen if this latest revolt within the veterinary profession will influence BVA support for badger culling.

The longer the veterinary industry provides a shield to the government and the NFU for the continuation of culling the greater will be the damage to its reputation in the eyes of the public. The profession is rightly putting scientific innovation and animal welfare at the centre of its long term vision, but this is undermined by its failure to apply these principles to badger culling. Rather than being an obstacle to change, vets hold the key to finding more effective solutions to reducing bovine TB that protect not only the interests of farmers, but also the future of our precious wildlife.

17

COST OF THE CULL

When Defra officials developed the badger cull policy in 2010 they established a working group which included representatives from the National Farmers Union. The group's existence was not made public and its recommendations were kept confidential.

A key priority for the group was to develop a risk register for ministers which weighed up the key benefits and disadvantages of moving forward with a badger culling policy. The existence of this register only came to the public's attention after a two-year legal battle between the Badger Trust and Defra, which forced its release under Freedom of Information legislation.

The Badger Trust requested the risk register in 2012. Defra refused and the Badger Trust took up the issue with the Information Commissioner, who also supported the release of the information on public interest grounds

This resulted in both the Badger Trust and Information Commissioner taking Defra to the High Court to release the risk register. When the hearing was held in 2014 the judge found in favour of the Badger Trust and Information Commissioner and he demanded that Defra make full disclosure. Defra put the register on its website a few days before Christmas in December 2014 without any publicity.

The register showed that ministers were warned of the severe risks posed by the controversial badger cull in a wide range of areas. These included strong public opposition and a possible legal challenge, a significant risk that badger culling could increase TB in cattle due to increased risk of perturbation, problems with the scientific evidence base to support culling, and concerns about the humanness of free shooting.

Another key concern raised with ministers was the danger that the costs of the policy could rapidly escalate to a point where they exhausted the funds available to Defra. This was a particularly important concern in view of the fact that the Conservative-led coalition came into office in 2010 with a plan to make the biggest cuts in public spending in decades. On 20 October 2010, the Chancellor of the Exchequer, George Osborne, announced spending cuts of £81 billion across government, telling MPs: 'Today is the day when Britain steps back from the brink, when we confront the bills from a decade of debt.'

Despite David Cameron's claim to want to be the 'greenest government ever' Defra's spending was slashed. The Defra Secretary of State, Caroline Spelman, believed she would benefit politically from offering up big cuts early in the Whitehall Spending Review process. This resulted in Defra's budget to 2015 being cut by 30%, and 8,000 job losses from a 30,000-strong workforce.

Against these savage cuts alarm bells soon started to ring in the department over the costs of moving forward with badger culling.

The policy was developed as a farmer-led policy. Free shooting was to be piloted as a killing method, as it would allow the NFU to establish culling companies and recruit local pest controllers who could be licensed to kill badgers with the minimum of supervision and training.

However from day one the costs to the taxpayer were completely under-estimated and largely hidden.

At no point was much thought given to the likely failure of free shooting as a killing method despite the fact that many experts who shot deer and other species for a living raised concerns about the ability of marksmen to kill a muscular low squat nervous animal like a badger at night.

In a report to Parliament before the culls started Defra confirmed that free shooting was estimated to cost only £300 per square kilometre compared to £2,500 for trapping and shooting — which showed the huge cost if free shooting proved ineffective. Most of the higher costs of trapping and shooting compared to free shooting, were due to the need for specially-trained government officials and cages to be brought into the cull zones in place of low cost cull contractors.

Another key cost factor that was completely underestimated was the cost of policing the cull. By allowing the NFU to create a climate of fear around the badger cull protest movement, the police overreacted and developed heavy handed policing strategies which soon ran into millions of pounds, passed back for Defra to pay.

In view of the public anger generated by the cull policy Defra also found itself flooded with requests for information under Freedom of Information legislation. Every aspect of the cull from its scientific justification, to its humanness and overall cost to the taxpayer was being dissected by thousands of people who were then flooding Defra with FoI requests.

The Badger Trust also tied Defra in knots in the courts with numerous legal challenges under FoI legislation and a major legal challenge in the High Court in 2014 on the failure to maintain the Independent Expert Panel to monitor humaneness for the second year of the pilot culls, which added to the public costs of the policy.

Compliance and monitoring costs also soon started to add up, with £2.6 million spent on humaneness and efficacy monitoring in the first year of the cull alone. Although the Independent Expert Panel was disbanded for the second year, wider monitoring, training and compliance costs continued to rise.

By the end of 2013 Defra which had seen its operating budget cut by over £900 million and was laying off hundreds of staff, confirmed that in 2011/12 it had spent £6.3 million of its £2.2 billion budget killing 1,879 badgers in two small areas of Gloucestershire and Somerset – which worked out £3,350 per badger.

Suddenly a policy which Caroline Spelman and her successor Owen Paterson had assured MPs would largely be funded by farmers was costing the taxpayer millions and had become the most expensive wildlife cull of its kind.

In 2013/14, based on its own figures, Defra spent £5,000 killing the American bullfrog, £15,000 killing non-native mussels, £45,000 killing non-native cray fish, £46,000 killing monk parakeets £120,000 killing the ruddy duck, and £10 million killing badgers.

Defra officials defended the rising costs by saying they needed to be seen in the context of the threat of bovine TB to the cattle industry which they estimated could cost the taxpayer £500 million over the next decade.

Defra has often been in denial about the financial burden of the cull. On many occasions in 2014 after the second year of culling, I was quoted in the media as saying the cost to public purse had already exceeded £10 million. Owen Paterson and his officials claimed such estimates were grossly exaggerated and part of a wider campaign by anti cull campaigners to undermine public confidence in the policy. Nevertheless my figures proved to be correct. In late August 2015 Defra released a full breakdown of the costs of the first two years of the cull. Including a one-off £2.5 million for delaying the start of the cull in 2012, the total costs were £16.7 million.

All this expense for what was supposed to be a farmer-led and funded programme. In any other area of government expenditure such costs would have led to an urgent cost benefit review but this was not the case for badger culling.

Indeed the cull was extended again. With the Conservatives now freed from the shackles of a coalition following the 2015 general election, Liz Truss announced in August that the cull would go ahead for a third year and be extended into Dorset. She used her speech to the Conservative Party's autumn conference to make a commitment to extend the cull to other areas. Following a weakening of the requirements for a culling licence, by early 2016 Natural England received 29 expressions of interest from farmers to cull in new areas, including Devon, Cornwall, Wiltshire, and Cheshire.

Including policing, licensing and monitoring and Whitehall costs, extending the cull to ten or more areas over the five years, could cost the taxpayer more than £100 million.

The Environment Food and Rural Affairs (Efra) Committee has now called on Defra to establish a thorough evidence base for its TB reduction policy and to communicate this in fully transparent manner. Meeting this demand will be no easy task as Defra mandarins have not so far provided any reliable evidence to show that the badger culls have lowered TB in cattle in the cull zones.

At the time of writing, in June 2016, the Defra Chief Scientist and Vet are both on the public record saying that it far too early to see any impact from culling and it could be eight years before any statistical trends link culling to any fall in bovine TB. They also stated that it will be difficult to assess the impact of culling alongside cattle measures.

With none of the badgers killed being tested for TB the hugely expensive cull is a bafflingly blind and indiscriminate process. TB data is also of little help as it is not broken down into the geographical detail needed to assess the impact of the culls.

From what we can ascertain based on the latest available Defra TB data, TB infection rates in cattle in Gloucestershire have seen no significant change between 2014 to 2015 whilst in Somerset TB incidents have gone up from 297 to 320, a 7% increase.

Defra's figures for England to the end of February 2016 showed the number of cattle slaughtered in the high risk areas, including the culling zones of Gloucestershire, Somerset and Dorset, increased by 11% compared to the same period in 2015.

The number of new herd incident in high risk areas increased by 5% compared to same period in 2015. This is despite expenditure of £25 million killing 4,000 badgers.

Defra has provided no breakdown of cattle TB rates in the cull zones despite being asked for this info by the Efra select committee.

The real tragedy of the huge waste of public money on killing badgers is that it does nothing significant to rid herds of bovine TB. With TB skin tests missing one in two infected cattle, the money to kill badgers would have been far better used – and at the time of writing would still be better used – to provide farmers with more accurate gamma interferon blood tests for TB, backed up by more widespread pre- and post-movement cattle testing controls.

Badgers are also paying a very heavy price for the increasing controversy over the rising cost of the cull. In April 2015 after much public pressure and internal discontent from its members the British Veterinary Association announced that it was withdrawing support for the use of free shooting as a culling method on humaneness grounds. The BVA stated that the monitoring results from the first two years of culling had not demonstrated conclusively that controlled shooting could be carried out effectively and humanely based on the criteria set for the pilots.

BVA President John Blackwell said:

> 'BVA's support for badger culling as part of the bovine TB eradication strategy has always been predicated on it being delivered humanely, effectively and safely. BVA supported the pilots to test the use of controlled shooting, but data from the first two years of culling has not demonstrated conclusively that controlled shooting can be carried out effectively and humanely based on the criteria that were set.'

In normal circumstances such a statement from the leading veterinary industry trade body would have brought an end to use the free shooting for all future culling operations. However the NFU quickly made it clear to Liz Truss and senior officials at Defra that any move to stop using this killing method, would make the costs of the cull politically unsustainable.

With already nervous officials at Defra saying that the costs of the badger cull were getting out of control the farming minister George Eustice was despatched to inform the media that the BVA's demands to end free shooting would be rejected. Eustice claimed that free shooting badgers at night was no different to shooting deer and foxes and that the animal welfare concerns raised by the BVA were not supported by the evidence Defra had received as part of its cull monitoring activities.

On learning of Eustice's response, I gave an interview to *The Times* confirming that in my view Defra had rejected the recommendation on cost grounds, and at the request of the NFU. Minette Batters, Vice President of the NFU, quickly responded, denying that NFU was trying to kill badgers on the cheap and defending free shooting on the same grounds as George Eustice.

The BVA were shocked to see its recommendation to stop the use of free shooting rejected so quickly by Defra and the NFU, and was left in the unenviable position of disagreeing with the method of killing badgers, but supporting the cull.

What the row over free shooting exposed was the extent of the NFU's influence over the cull and the government's concern about anything that might increase its cost. The lack of openness and transparency over the policy is quite remarkable. Up to the end of 2015 the government refused to release its cost benefit analysis for the cull. When it finally did publish the document, *Bovine TB: Control Policy Value for Money Analysis*, on 17 December 2015, it was so heavily redacted it was all but useless.

The NFU has fiercely guarded information on the cost burden being carried by the farming industry for the cull policy, claiming it is commercially confidential, and its release could put farmers and cull contractors at the risk of threats and intimidation from anti cull protesters.

Parliament has also failed in its duty to hold Defra to account for the huge waste of public money on killing badgers. With Defra cutting budgets for key areas of work including flood defence, pollution control, and environmental stewardship, the Efra and Treasury select committees should be asking some very hard questions of the Environment Secretary and NFU over the £100 million that could be wasted on killing badgers by 2020.

18

FARMING FUTURE

Many of us still have a romantic vision of farming based on children's books with their bright pictures of farms, made up of green fields, hills, trees and hedges. In this idyllic landscape, happy chickens, cows and sheep roam, while a farmer and his wife — holding a basket full of fresh produce and smiling warmly — stand beside their trusty red tractor. We all remember our harvest festivals in school, where we brought in vegetables and fruit, and discussed how they were grown and celebrated the wonder of nature.

The major supermarkets continue to play on these themes in their marketing campaigns with big murals and images of bright beautiful landscapes in their stores, celebrating the best of British farm produce. They even create fictional farm brands, which mislead their customers into thinking they are buying produce from an individual farm, when in fact it is sourced from a multitude of production facilities from around the country or, in some the cases, the world.

Flagship rural programmes like the BBC's *Countryfile* mostly focus on the niche high-value farming sector, which produces the best quality fresh produce, dairy and meat products in beautiful landscapes and to the highest production and animal welfare standards.

We are constantly told by the government and the farming industry that farmers are the guardians of the countryside, wildlife and the natural environment and that without them our rural landscapes would be a wild barren wilderness. The reality could not be further from the truth. Farming in 21st century Britain is a highly-intensive industry and the fate of the badger and much of our wildlife is now directly linked to the industrialisation of agriculture and food production.

Today the UK has over 142,000 farm businesses, which is more than the number of companies involved in the motor trade, education, finance, and insurance. These farming businesses are the cornerstone of UK food and drink, which is valued at 16% of total UK manufacturing turnover. Sales of UK food and drink were £82 billion – with exports of £12.8 billion – in 2014, according to the Food and Drink Federation. When combined with farming, the sector employs 3.8 million people.

When you consider these figures, you begin to understand the clout the NFU has with Defra's civil servants and politicians.

The NFU was established to represent the interests of farmers at the Smithfield Show in 1908, at a time when farming was going through a deep depression as imports of cheap grain and frozen meat drove down farmgate prices.

By 1914 it had started to establish its credentials as the voice of the British farming industry and was drawn into agricultural policy discussions at a national level. The threat to food production as a result of heavy losses of merchant shipping to German U-boats during the first world war rapidly increased its power and influence in government.

In the 1920s and 30s it tried to become a political force by running candidates for Parliament or sponsoring Conservative candidates that supported its views on farming issues. However the NFU soon found that it did not need MPs in Parliament to influence the political process: in the run up to the second world war the Ministry of Agriculture, Fisheries and Food (Maff) became increasingly reliant on the NFU to maintain food security.

In the post war period, the relationship with Maff was reinforced by the 1947 Agriculture Act which effectively set in law the involvement of the NFU in all key aspects of agricultural policy development including agricultural subsidies and prices paid to farmers.

For the next 65 years the NFU continued to grow in power and influence by shunning the more militant strike tactics employed by other trade unions, in favour of maintaining a close mutually beneficial relationship with Labour and Conservative governments.

In many ways the NFU acts as a government agency when it comes to agriculture and farming policy. Its officials are regularly consulted on policy issues and often work behind closed doors with civil servants on sensitive issues.

It is a professionally-run run organisation with very effective lobbying, campaigning and media operations.

It might only represent around 15% of all the farmers in the UK but most major farm businesses and landowners are within its ranks and it has established a close relationship with other key landowning and food organisations to strengthen its influence.

It also established a very successful NFU Mutual Insurance Company in 1910 which today ranks among Britain's 10 biggest insurers. Although this is now run as an independent business from the NFU, it maintains close links to local NFU group secretaries and directly funds the organisation. Indeed the competitive insurance rates provided by NFU Mutual motivate many farmers to retain NFU membership.

The NFU calls itself the voice of British farming, but in many ways it is more the voice of modern intensive agriculture. It sees farming as an industry where production and profits are the most important issues. All its policies on disease control, pesticide use, environmental stewardship and wildlife protection are driven by this overtly commercial agenda. This is despite the fact that farming covers 69% of the land in England and the way in which it is managed to produce food, fuel and fibre has a huge impact on the natural environment, the survival of wildlife and our own health and well-being.

The tragedy is that the NFU view on farming and food production is largely shared across the food chain and by policymakers and politicians.

Today Defra is not a Department of State to protect the environment. Just like its predecessor Maff, it has become an arm of government to protect business from the environment.

Its primary aim is to increase food production, productivity and profits, just like the NFU its policies largely ignore the vital role our landscapes used for farming play in terms of wildlife and habitat protection.

To see the negative impact of the profits and productivity led agenda on both farmers and wildlife, we need only look at UK dairy farming. Today the industry is in crisis with many supermarkets selling milk cheaper than bottled water and dairy processors continuing to cut prices to farmers. It costs on average 30p a litre to produce milk, but many farmers are paid 20p a litre. Half of Britain's dairy farms have gone out of business in a decade, mostly small family farms.

The origins of the crisis in the dairy industry go back to the abolition of the Milk Marketing Board (MMB). The MMB was established in the 1930s to guarantee a fair price for milk producers, regardless of the internal market. This provided an important safeguard for small farms from the power of the milk processors and supermarkets and also gave the next generation of farmers the security they needed to enter the dairy industry.

However, it was widely disliked by the larger dairy farms who believed it prevented their expansion by keeping unprofitable farms in business. This view was largely shared by the NFU, which welcomed the MMB's abolition in 1994.

With the abolition of the MMB and no mechanism in place to protect small dairy producers, the farmgate price of milk fell by 28% between 1994 and 2010. Farmers in remote rural areas also found tanker collection costs increased.

As smaller dairy farmers went out of business, larger and more intensive farms grew in number. In 1996 the average dairy herd was 78 cows. By 2013 more than one in ten dairy herds had more than 500 cows. We are also seeing an increase in mega dairies that have more than 1,000 cattle.

Many of these large dairy herds are kept indoors for significant periods of the year, in some cases never being allowed to graze outside. Cows in these intensive indoor systems are more likely to suffer from stress and disease and produce huge amounts of slurry emitting more greenhouse gasses and polluting water systems.

In July 2009 the European Food Safety Authority addressed the impact of keeping cattle indoors for most of the year in intensive cattle systems in a report titled, *Effects of farming systems on dairy cows welfare and disease*. It raised concerns that high yielding cows lost excessive body condition in the early stages of lactation – 'milked to starvation' as it has been called in the media. Increased in-breeding of cattle associated with reproductive problems, negative energy balance, and increased risk of infection diseases were other problems. All these factors put cattle at risk of increased spread of bovine TB.

In its guidance to farmers (*Bovine TB: Guidance on the management of public health consequences of tuberculosis in cattle and other animals (England)*), published in September 2014 Defra stated that transmission between cattle occurs most commonly by breath. Such risks are increased by packing cows close together.

Another problem is slurry. A review of the potential role of cattle slurry in the spread of bovine TB by the Agri-food and Biosciences Institute in February 2014 found that slurry can contain *M.bovis* and contaminate pasture soil and silage. The research also found that spreading slurry can generate aerosols that potentially carry TB bacteria for considerable distances.

The intensification of the dairy industry has also brought an increasing reliance on exports. Britain is only 81% self-sufficient in milk production, but with deregulated milk prices and access to the global market, many food retailers are importing milk from the cheapest supplier they can find. In 2013 the UK imported over 132 million litres of milk, forcing British farmgate prices down even further.

Rather than sell milk to British consumers, the fall in prices is forcing many farmers to rely on export markets in countries such as China and Russia. However in 2014 Russia banned imports of milk from the EU in reaction to criticism of its behaviour in Ukraine. Also China, which now has a stockpile of milk powder, is reducing milk imports and plans to establish its own 100,000-cow mega dairy to supply Russia with milk.

With the EU also abolishing its milk quotas in 2015, which were aimed at preventing over-production by farmers, the export market for British dairy farmers is looking increasingly bleak.

The reaction of the government and the NFU to the crisis shows the danger of their reliance on their export-led vision of the future.

The Environment Secretary, Liz Truss, belatedly called for improved labelling to promote British dairy products in supermarkets, but refused to get tough with the major retailers by extending the remit of the largely toothless Grocery Code Adjudicator to provide better protection for dairy farmers. The European Commission came up with a £26 million aid package for UK dairy farmers as part of a wider EU-wide assistance programme for the sector, but even this was geared towards larger farms producing higher volumes of milk.

The NFU also found itself conflicted when dealing with the crisis. It only started to react when groups such as Farmers for Action started to blockade supermarkets, in some cases marching their dairy cows through them. It pushed Defra and the major supermarkets to put in place a minimum price for milk, but this went against its long term policy of calling for an end to milk quotas and greater trade liberalisation in the sector that benefits large intensive dairy farms.

The crisis in the dairy industry is of crucial importance to the debate over badger culling as the industry is on the frontline in the battle against bovine TB. With over 95% of bovine TB infection being cattle to cattle, the rapid intensification of dairy herds and the fact they are being kept indoors for longer periods of time is a major risk factor in the spread of the disease.

As dairy farms face collapse due to falling milk prices, farmers are increasingly supplementing their income by moving and selling calves and cows, often with poor biosecurity, testing and movement controls, which can increase the spread of bovine TB.

Another often overlooked factor in the dairy industry crisis is its impact on the issue of TB cattle vaccination.

Trials in Ethiopia and Mexico showed a TB vaccine could be 58-68% effective in preventing the spread of the disease.

For the past five years the government has stated it will trial a TB cattle vaccine in the UK. In 2014 Defra commissioned a consortium including Triveritas UK, scientists from the Animal Health and Veterinary Laboratories Agency and Cambridge University to design field trials in the UK. Triveritas, which specialises in undertaking livestock field trials, was to design the trial for the vaccine together with a new diagnostic test for differentiating between infected and immunised cows.

Despite the importance of developing an effective TB cattle vaccine, Defra announced in 2015 that it had called off the trial on cost grounds. This decision angered many farmers, but the NFU was largely silent on the issue. In fact, many in the NFU leadership were happy to see the project kicked into the long grass, in view of the potential impact of vaccines on the export of cattle, meat and dairy products to EU countries and important wider markets like China.

With an ever increasing reliance on exports for the British livestock and dairy industry, the NFU lives in constant fear of the doors to key markets like China closing as a result of a food safety ban on products from TB vaccinated cattle.

The crisis in the dairy industry is a perfect example of all that is wrong with the modern farming industry and how the NFU's influence over government policy has failed most farmers, taxpayers and wildlife.

Dairy farmers are being forced to make decisions to survive which are increasing the spread of bovine TB. From larger herd sizes, more frequent cattle movements, and greater reliance on exports, the industry is caught in a vicious circle, which will ultimately lead to the death of more badgers, who through no fault of their own are paying the ultimate price for the failings of modern farming.

19

FATE OF THE BADGER

In 1986, a naturalist, writer and biological researcher Richard Meyer wrote a book called *The Fate of the Badger* to draw public attention to the persecution of the badger and a disastrous badger cull then being implemented by the Ministry of Agriculture

The world has changed a great deal in the 30 years since he wrote his ground-breaking book, but not when it comes to the debate about bovine TB and the role of the badger in spreading the disease. Every issue in *The Fate of the Badger* is as relevant today as when it was published. The lies and propaganda of the farming industry, the politicisation of the science, the demonisation of a species for short term political and economic interests, nothing has really changed.

Richard Meyer woke up the nation all those years ago to what we were doing to our precious largest surviving carnivore which has lived in our isles for half a million years. For the first time he laid bare the brutal reality of how the badger had become the innocent victim of a disease which was spread by modern intensive farming practices and poor control measures.

Sometimes it takes a more detached observer to see what is happening. In February 2013 senior officials from the European Commission gave evidence to the Environment Food and Rural Affairs select committee (Efra) in Westminster on the UK's TB eradication policy. Bernard Van Goethem, the Commission's Director for Veterinary Affairs, and Francisco Reviriego, its Head of Disease Control and Identification, expressed their serious concerns about the direction of bovine TB policy in the UK.

They started their presentation by providing MPs with a map of Europe showing TB status in EU member states. The vast majority of the map was green, showing member states with official TB free status. A number of countries including Spain, Portugal, Italy, Greece, and Romania were in a low risk category with occasional isolated outbreaks of bovine TB, but the key focus was on the UK and Ireland which had the worst bovine TB status in Europe.

The officials told MPs that the scale of the UK's bovine TB was without precedent in the EU. Van Goethem laid the blame for the problem at failures with the government and the farming industry. He claimed that due to cattle export restrictions at the time of the BSE crisis in the 1990s, farmers had no incentive to put in place strict TB controls to gain disease free status. He told MPs the situation was made worse in the aftermath of the foot and mouth crisis in 2001, when hundreds of thousands of cattle were moved from TB hot spot areas without any TB testing controls for almost two years.

He also raised serious concerns that the UK was moving millions of cattle ever year with poor biosecurity controls, which was leading to the spread of bovine TB. When pressed by MPs on the badger issue, he accepted that wildlife could be a factor in the spread of the disease, but emphasised that other EU member states had achieved disease free status through TB testing, movement controls, and, where necessary, whole herd depopulation – not through mass culls of badgers or other wildlife.

MPs pressed the European Commission officials on the use of a TB cattle vaccine to slow the spread of bovine TB. Van Goethem confirmed that field trials would need to take place in the UK to prove a vaccine could be effective and the EU had laid out a 10-year timetable with the UK government to enable this process to go forward and allow the export of vaccinated cattle to the EU.

He ended his presentation by warning the government and the farming industry to take much tougher action on tackling bovine TB in cattle. He confirmed that the EU had a total annual disease control budget of 200 million euros and that over 30 million of this budget would be allocated to the UK annually to help reduce the spread of bovine TB. However in return for this significant investment from its disease control budget, the European Commission would need to see significant improvements in TB disease control measures in the UK including tighter TB testing systems, movement controls and steps to reduce the level of cattle trading.

The Commission was to be applauded for telling MPs some of the inconvenient truths on the issue of bovine TB control and they clearly believed that a carrot and stick approach in terms of delivering EU funds could bring about significant improvements in terms of reducing the spread of the disease.

In the three years that have passed since the Efra select committee hearings, over 90 million euros has been pumped into the UK TB eradication strategy. This has resulted in some tightening of cattle movement and TB testing controls, but with the latest data from Defra showing a small increase in TB rates in cattle in both high and low risk areas, the European Commission has to date little show for its 90 million euro investment.

Three years on from his presentation to the Efra committee in Westminster, I asked Van Goethem, now acting head of the European Commission's Health and Food Safety Directorate, if much had changed on bovine TB in the UK. Showing clear frustration with the lack of progress, he said he welcomed the UK's TB eradication strategy, but much remained to be done if bovine TB was to brought under control. He joked that cattle movements were a national sport in Britain, but I could sense his frustration in the failure of the government or farming industry to bring about any significant reduction in the level of cattle trading.

He would not be drawn on the rights and wrongs of the badger cull, but he emphasised that no EU funds were being used for killing badgers. As with his presentation to MPs in 2013, he gave me a clear impression that the European Commission found the UK government's obsession with killing badgers a dangerous distraction from tackling the root cause of bovine TB in cattle.

When you remove the badger from the table, very little divides wildlife conservationists from farmers and policy-makers when it comes to agreeing a strategy to reduce the spread of the disease.

We have no solid evidence to prove that badgers pass TB to cattle and that killing them can makes any significant contribution to lowering bovine TB. On the other hand we have a mountain of scientific data and proven field research to show that bovine TB can easily spreads within the cattle population and that cattle based measures are key to eradicating the disease.

After 40 wasted years no one should be in any doubt that culling badgers is a hugely costly, scientifically-flawed and cruel wildlife destruction policy, that is a dangerous distraction from getting on top of bovine TB in cattle.

The tragedy of this debate is that badger protection campaigners do want to prevent the spread of TB in badgers and cattle. For 40 years we have been telling politicians, policymakers, NFU, and the BVA that they must stop playing the badger blame game for short-term political and economic gain and focus on finding long-term cattle-based solutions to the problem. We all know the solution to bovine TB control comes down to tight TB testing systems, movement and biosecurity controls and herd depopulation where the disease is persistent in livestock We could kill every badger in Britain but this reality would remain unchanged.

We know this, not just from the evidence in scientific journals, but from an experiment in another part of the United Kingdom, Wales. In Wales, too, there was a clamour for a cull of badgers but, following a successful legal challenge by the Badger Trust in 2012, a proposed cull there was abandoned. Instead for the past years, the Labour administration in Cardiff has put in place an annual TB testing system for cattle, supported by tighter movement and biosecurity controls, along with a badger vaccination trial. The result? The number of cattle slaughtered for TB in Wales has fallen by 45% in the past five years. Today 94% of cattle in Wales are TB-free.

The simple fact is that badgers got this disease from cattle as a form of industrial pollution from intensive farming systems.

We have no right to risk local extinction of the species from areas of Britain where they have lived for 500,000 years because of incompetence, negligence and deceit in the government, farming and veterinary industry.

Sadly it will not be the issue of scientific validity or humaneness that finally brings an end to the cruel madness of killing badgers. As with most political issues, it will be the cost in pounds or euros.

As Defra finds its room for manoeuvre increasingly restricted by savage budget cuts, it knows it cannot deliver on the NFU's ambition of a nationwide badger cull. In these circumstances the EU's annual 30 million euro lifeline towards the cost of bovine TB reduction, becomes all the more important for Defra and the Treasury. However the political and economic earthquake caused by the Brexit vote in the EU referendum on the 23 June 2016, is likely to see this funding cut off very quickly leaving a budget black hole when it comes to the future management of bovine TB in England.

Farmers in the UK are also becoming increasingly frustrated by the failure of the government and the NFU to help them reduce bovine TB. Rather than the endless battle over the killing of badgers, which is doing so much damage to their reputations. many are now openly calling for an alternative strategy. They rightly want to know why they are not being offered free access to gamma interferon testing, when so much public money is being wasted on killing badgers.

They are becoming increasingly angry by the government and NFU's reluctance to push for TB cattle vaccines to control the disease, to keep open the cruel and highly controversial live animal export trade in Europe and beyond. They are sick to the back teeth of the failure of the NFU and the government to stand up to the major food retailers, who have driven down the price of milk and put thousands of dairy farmers out of business.

They are concerned by the failures in the state and private veterinary sector to ensure TB tests are carried out effectively and accurately, which is resulting in the destruction of cattle later found to be TB free.

They feel increasingly uncomfortable with the government selling TB meat into schools and hospitals, without any labelling or traceability in order to generate millions of pounds to keep the TB cattle compensation scheme going.

Writing the foreword for the 30th anniversary edition of Richard Meyer's book, *The Fate of the Badger,* I realised that if our political leaders and the farming industry had taken note of his wise insight in 1986 we would not be seeing tens of millions of pounds wasted on killing badgers, in a cruel culling policy which has no scientific or animal welfare justification.

Three decades on, as I come to an end of my book about a new generation who are making a stand to protect badgers, the question remains: will we keep repeating the mistakes of the past or can a solution be found to bovine TB which does not come at a huge cost to the badger.

After decades of incompetence, negligence and deceit will the politicians and farming lobby finally wake up to the true cost of the badger blame game to farmers, tax payers and the future of our precious wildlife.

If I come to write a 30th anniversary version of *Badgered to Death* in 2046, will badgers have disappeared from many of our landscapes forever, having paid the ultimate price for our ignorance, greed, and stupidity?

Dominic Dyer
June 2016

BADGER VOICES

Most authors acknowledge family members or other individuals who have given support or assistance in the writing of a book. For *Badgered to Death* I wanted to go much further than simply name-checking people. The journey I have taken in defence of the badger would have been very lonely and difficult without the inspiration and support of many compassionate people along the way.

Some of these people are well known conservationists or public figures, but many are largely unsung heroes whose contribution to protecting the badger in the face of unparalleled demonisation and destruction will go largely unnoticed in the history of the conservation movement.

Badger Voices is my opportunity to ensure they receive some well-deserved credit for standing up for badgers and being true stewards of the natural world for future generations.

CHRIS PACKHAM

After Sir David Attenborough, Chris Packham is Britain's most respected and high profile naturalist and broadcaster. Many people who achieve such success and public acclaim protect their image and status to the point of never entering into areas of political controversy.

For Chris, this has never been an option when it comes to standing up for badgers. Despite repeated attempts by the BBC to silence him, he has spoken of his profound anger at the dishonesty, brutality and manipulation of science by the government and farming lobby to justify badger culling. Rather than hiding away from protests against the cull, he has given his strong personal backing and proudly stood alongside the Badger Army when it came to Winchester in November 2014. He has used his influence within the BBC to ensure both *Springwatch* and *Autumnwatch* have provided a balanced view on the badger and bovine TB debate and has worked tirelessly to promote a positive image of the badger as one of the nation's most iconic wild species.

He has even been willing to shame some of Britain's largest conservation charities for taking a vow of silence on politically controversial issues such as badger culling, fox hunting, and the plight of hen harriers. His talk of 'fence-sitting and ineffectual risk-avoidance' struck a chord with millions of wildlife lovers across the country and in the case of the Wildlife Trusts has undoubtedly had a significant influence on forcing them to take a stronger line of opposition to the badger cull.

Despite repeated attempts by certain section of the media and the outspoken chief executive of the Countryside Alliance, Tim Bonner, to have him removed as a BBC broadcaster, Chris has become even more popular with wildlife lovers across the country as a result of his willingness to fight political battles for badgers, foxes and hen harriers and his stock as a leading BBC broadcaster has risen to new highs.

Without Chris taking me aside at the Birders Against Wildlife Crime Conference in Buxton in May 2015 and urging me to tell this story to the wider world, this book would not have been written.

His willingness to write its foreword and to put his name and credibility behind it despite the controversial issues I have covered, including the failure of his principal employer, the BBC, tells you that Chris Packham is not just another nature broadcaster providing armchair entertainment. He is willing to challenge the political, business and media establishment and even ruffle feathers in the conservation world to protect wildlife.

BRIAN MAY

I first met Dr Brian May in a Starbucks next to St James Park Station in September 2012 prior to meeting the head of the National Trust to discuss the cull. Having grown up with Queen as one of the most successful rock bands in the world, I was not quite certain what to expect. To my pleasant surprise I found Brian to be a highly intelligent man with a deep interest in the science of bovine TB and a genuine commitment to protect badgers.

He not only funds the rescue and rehabilitation of badgers and other British wildlife through his Save Me charity, but he has also played a key role in bringing wildlife charities together through his Team Badger campaign group to put political pressure on the government to stop the cull policy.

He has led debates in Parliament and spoken at numerous public campaign events across the country on why the badger cull has no scientific or animal welfare justification. He has taken on the government and farming lobby on behalf of the badger protection movement.

He has taken the bold step of visiting dairy farms to debate with farmers on bovine TB and badgers and has even joined wounded badger patrols in the cull zones to show his solidarity with wildlife campaigners on the frontline of badger protection.

At times he has come under quite vicious attack from organisations like the Countryside Alliance and pro-cull farmers and livestock vets, but he has not wavered in his determination to stand up for badgers and other British wildlife.

As a result Brian is today one of the leading figures in the animal welfare and wildlife protection movement in Britain. He is on the record as saying he would rather be remembered for his work protecting animals than as one of the world's leading rock guitarists.

Brian is far more than a rock star with a cause, he is an inspiration and guiding light for wildlife defenders in Britain and around the world and he deserves his place in the wildlife hall of fame as well as the rock and roll hall of fame.

BILL ODDIE

As a child growing up in the 1970's, I had fond memories of watching Bill Oddie in *The Goodies* or appearing on *Top of the Pops* singing the 'Funky Gibbon' and 'Black Pudding Bertha.'

Watching Bill as a comedian and performer all those years ago, I never thought that nearly 40 years later I would be standing alongside him fighting to protect badgers outside of the High Court and at Badger Army protest marches across the country.

In the post-*Goodies* era from the 1980s onwards Bill carved out a new and very successful role as a nature broadcaster and writer. He reached millions of people as a presenter on *Britain Goes Wild* and *Springwatch* and *Autumnwatch* and won over a new generation of viewers with his unique broadcasting style which combined humour with a deep understanding and respect for wildlife.

Bill has never been afraid to enter into political controversy to stand up for wildlife from fighting to protect whales in Iceland to taking on the powerful shooting lobby in Malta to prevent the mass killing of birds. Through his work with numerous wildlife protection charities and rescue centres in Britain, he is never far away from the heated debates about fox hunting, use of snares and badger culling.

Bill's commitment to the badger protection cause has been outstanding. He has spoken at anti badger cull marches in Bristol and Birmingham, been a regular participant in wounded badger patrols in Gloucestershire and Somerset and helped draw media attention to the Badger Trust legal challenge against the cull in 2014, by joining me and hundreds of anti cull protesters on the steps of the High Court.

When I was appointed chief executive of the Badger Trust, Bill was quoted as saying: 'Dominic Dyer is the man you want speaking on your side especially if you're a badger.'

To many people Bill Oddie might remain a defining figure of 1970s TV comedy, but to me he has a far more important legacy as a committed wildlife champion and badger protector, who is willing to stand up for what he believes.

SIMON KING

Simon King is one of Britain's most respected nature documentary-makers and broadcasters. In a career that has spanned over 30 years he has worked on all the major BBC natural history series including *Wild Africa, Blue Planet* and *Life of Mammals.* He became a household name for his work on *Big Cat Diary* and *Springwatch* and *Autumnwatch.*

As President of the Wildlife Trusts Simon has been a strong voice of opposition to the badger cull policy. He has a good understanding of the science of bovine TB and has been very careful not to alienate farmers in his campaign to protect badgers.

Simon has eloquently argued for the defence of the badger in debates in Westminster with the Badger Army on the streets of Bristol and town hall meetings in Cheshire. He joined me for an Oxford Union-style debate on the badger cull at Bristol University on 16 May 2014, alongside the Vice President of the NFU, Adam Quinney, and Dr Lewis Thomas of the Veterinary Association for Wildlife Management. A vast majority of students who attended the debate voted against badger cull, which is testament to the strength of Simon's scientific and ethical arguments against the policy.

Simon has also brought the wonders of badgers and other British wildlife into our homes through his live camera network at his 10-acre plot known as Wild Meadows and has now set up a charity to reclaim more land for wildlife protection.

In a debate which often generates a great deal of anger and emotion, Simon's calm thoughtful science based approach has gained him much respect in both the wildlife conservation and farming community.

PATRICK BARKHAM

Patrick Barkham is one of Britain's leading wildlife writers and environment journalists. His book *Badgerlands* has inspired a new generation to understand the importance of the badger to our national heritage, culture and landscapes. Published in June 2014, in advance of the second year of the badger cull, *Badgerlands* is a beautifully written and engaging book, which takes the reader on a journey of discovery about this shy enigmatic mammal, which splits public opinion and generates such political controversy.

As a result of the success of *Badgerlands* and his role as a natural history writer for *The Guardian*, Patrick has played a key role in keeping the badger cull debate in the public eye. He is one of the few journalists who has the taken the time to get to know and understand the badger protection movement and its wider impact on the wildlife conservation debate in Britain today.

Patrick was the first journalist to uncover the close relationship between the police and the NFU on the sharing of information on protesters in the cull zones. He has also not been afraid of making the connection between the demonisation of badgers and the increased level of persecution of the species and has not held back from accusing the government of culling badgers to gain political support in the farming and wider rural community.

Like Chris Packham, Patrick Barkham has played a major role in the publication of this book. He encouraged me to tell my story about the politics behind the badger cull and gave me invaluable advice about starting out as author and how best to secure a publishing contract.

Patrick is a gentle thoughtful man with a steely commitment to wildlife protection. He uses his skill as writer and a journalist to both inspire wonder and respect for nature but also to generate anger to protect species such as the badger.

JONATHAN LEAKE

Jonathan Leake is Science and Environment Editor at the *Sunday Times*. Unlike many journalists he has maintained a strong focus on bovine TB and badgers over many years. In May 2013 he was the first journalist to get an in-depth interview with Owen Paterson, revealing how he was cherry-picking scientific evidence to justify the killing of up to 100,000 badgers in a nationwide badger cull.

Following *Channel 4 News'* refusal, it was Jonathan who got my TB meat story onto the front page of the *Sunday Times* in June 2013, shedding a light on the hypocrisy of the government over its claims that badgers should be killed to protect public health, when it was selling TB meat into hospitals and schools.

His refusal to be cowed on this issue is also testament to his independence as a journalist and his willingness to peel away the layers of government secrecy to get at a story of significant public interest.

Jonathan constantly keeps the Defra press office on its toes with his requests for key information on bovine TB policy and his digging has paid off with exclusive stories about Defra using captive badgers for TB testing research, the sacking of a Defra civil servant for criticising the NFU over the cull policy on social media and the doubts over the wisdom of the cull policy from leading scientists advising Defra and Natural England.

At a time when the quality of journalism is declining as fast as the circulation of newspapers, Jonathan remains one of the most hard working and respected science and environment writers in the national press. He has a strong commitment to wildlife protection and has used his influential and powerful position as a leading writer on the biggest-selling quality sunday newspaper to focus public attention on the scientific failures of the badger cull policy.

MARTIN HANCOX

Anyone involved in the public and political debate about the badger cull policy has probably been on the receiving end of emails from Martin Hancox. These usually come in the shape of bold capitals mixed randomly with lower case text, condensing complex scientific arguments against the badger cull policy with no consideration for grammar and often with outbursts about the stupidity of politicians, campaigners, and the media.

The sheer volume and intensity of these emails can lead many people to dismiss Martin as an angry eccentric, but this would be a big mistake. Martin is a zoologist and a former member of the government's Badger TB panel and one of the UK's leading experts on bovine TB and badgers.

Martin is sometimes dismissed by academics, policymakers and politicians due to his dishevelled appearance and difficulties communicating his knowledge and expertise to the wider world beyond his blizzard of emails. He remains on the fringes of the bovine TB debate, but anyone who takes the time to read his emails and meet the man in person soon realises that Martin Hancox has a huge amount of valuable knowledge and makes a great deal of sense on the bovine TB issue.

Martin rightly points the finger of blame for the spread of bovine TB on the cattle industry and it's good to see his intelligent science-based arguments against culling badgers gradually reaching a wider audience via letters and opinion articles in the regional press.

In a debate which is so heavily influenced by short term political and economic interests, Martin remains a breath of fresh air. He is hugely committed to protecting badgers but also helping the farming industry find a science based solution to bovine TB.

PETER MARTIN

I first met Peter Martin on a wounded badgers patrol in Gloucestershire cull zone in 2014. At the time Peter was helping Gloucestershire Against Badger Shooting with its media activities and had become a highly effective voice of the local badger protection movement.

Peter has a local government background as well as being an experienced trade publication editor and writer. With his upbringing and career, he could easily have been part of the rural shooting and hunting community, but he has chosen the path of compassionate conservation rather than killing for pleasure.

Putting his local government and media skills to good use he took on the role as chairman of the Badger Trust in 2015 at a critical time for the charity. With huge resolve and energy he has travelled up and down the country meeting badger groups, speaking at protest events and public meetings and taking a hands on role in supporting the trust's media and lobbying activities.

Peter recognises that the Badger Trust must remain at the forefront of the political and media debate on the badger cull issue, but he also works tirelessly to maintain strong links with badger protection groups around the country, which carry out critically important work rescuing injured badgers, dealing with badger persecution issues and planning developments that threaten badger setts.

Peter is not easily intimidated and has been willing to take on the government, farming and hunting lobby to defend wildlife. He has not shied away from controversy and led the criticism of the Somerset Wildlife Trust for appointing pro badger cull Michael Eavis and Rebecca Pow MP as vice presidents, raised concerns about the RSPCA softening its approach to wildlife and politics, and even accused the BBC of institutional bias over its coverage of the bovine TB and badgers issue.

Peter is a strong minded, passionate and articulate defender of wildlife, who leaves no one in any doubt about his resolve and commitment to protect badgers.

MARK JONES

I first met Mark Jones in 2009 when he started working for Care for the Wild International after returning to the UK following a number of years working as a vet for the Wildlife Friends Foundation in Thailand.

Mark is rare species in the wildlife conservation world, a qualified and highly-experienced vet who has committed his life to protecting wildlife life at home and abroad. Few people in the conservation world can match Mark for his intellect, drive and determination and ability to communicate a science-based conservation argument.

His role in the badger cull debate has been significant, when many vets have given in to peer pressure or put their financial dependency on the farming industry before their concerns about wildlife protection, Mark Jones has been a beacon of hope for care and compassion.

Mark has travelled up and down the country speaking out against the badger cull at marches, town hall meeting and debates with leading scientists, politicians and veterinary industry leaders. He has been a key figure in the media presenting a balanced intelligent science based argument against the badger cull and has not been afraid to put pressure on the Defra Chief Vet or the President of the British Veterinary Association over their continued support for the policy.

TOM LANGTON

Tom Langton is a successful ecologist with extensive field experience in wildlife protection and habitat management on a UK and international basis. He first came to my attention in 2014 as he mounted a single-handed campaign against the BBC over claims in David Gregory-Kumar's report that culling badgers in Ireland had lowered bovine TB in cattle.

Despite being initially rejected by the BBC editorial complaints unit, Tom persisted over a 12-month period with letters and phone calls to the BBC, until the complaint was elevated to the BBC Trust, which eventually agreed that the report was inaccurate and misleading.

Tom has vigorously pursued the government for not carrying out sufficient environmental impact studies on the culling of badgers through the UK courts. His drive and determination also led to the creation of EuroBadger, a unique coalition of badger protection groups from across Europe, which is already beginning to have an impact on the bovine TB debate in the European Parliament and the Commission.

Tom is not afraid to ruffle the scientific establishment in order to protect badgers and he has recently started to open up a much needed debate on flaws in the Randomised Badger Culling Trial.

DIANE BARTLETT

When historians look back at the wildlife protection movement in Britain in the second decade of the 21st century they will no doubt come across film footage recorded by Diane Bartlett.

As digital technology rapidly advances with 7 in 10 people now owning a smart phone in Britain, Diane has been at the cutting edge of amateur film making for the wildlife protection movement. Over the past three years as the Badger Army has marched in 35 towns and cities across England Diane has been on hand using the latest smartphone technology to capture the energy and excitement generated by this people power.

Diane's videos have been watched by large audiences in Britain and around the world and have played a vital role in generating public and media interest in the anti badger cull movement.

Diane has taken her video camera onto the streets, into the university lecture halls to the High Court and into the Labour Party conference to record speeches and debates on the badger cull issue. Using the latest technology along with her daughter Harriet, Diane has even been able to live-stream footage of marches and debates to audiences in UK and around the world.

Diane has caught all the key moments in the debate on bovine TB and badgers involving wildlife campaigners, politicians, scientists and vets and has helped to educate a new generation of people on the realities of intensive farming and wildlife protection in Britain today.

Whatever the weather or location Diane can be seen at the front of a march with her camera, capturing for history the care and compassion of the British public when it comes to protecting badgers.

Diane has a special place in the Badger Army and her work will live on for generations to come as a testament to what people can achieve when they stand together to protect wildlife.

SPARTACUS

Spartacus, as he is known on social media, has played a crucial role in the badger protection movement over the past four years. Using the power of social media and the Freedom of Information process, he has relentlessly pursued the government's dirty secrets on the badger cull.

It was Spartacus who helped me put the TB meat story on the front page of the *Sunday Times* and he was instrumental in using Freedom of Information requests to peel away the secrecy in Whitehall to get an accurate figure for the public cost of the culling policy.

His Twitter feed can send shivers down the spine of the Defra Secretary of State, Chief Vet, and the NFU as he pursues the incompetence and deceit which underpins the badger cull policy.

It was Spartacus that helped put me on the hill in Taunton in September 2013 to tell my story about why the badger cull was so wrong that set me on the path of writing this book. He might be unknown to many people beyond the faceless world of social media, but I am for ever grateful for his commitment and support in the fight to protect badgers.

EMILY LAWRENCE

I was first contacted by Emily Lawrence in September 2013 to say she would like to organise an anti cull protest march in David Cameron's constituency in Witney in Oxfordshire.

Emily was a pre-school teaching assistant with a passion for wildlife and a fire in her belly over the need to take the protest movement to David Cameron's front door.

Over the following weeks, she threw herself into organising the largest wildlife protest ever held in Witney, bringing hundreds of people together from across the country to march up the high street to the Prime Minister's constituency office and then onto Witney Green for speeches.

Following the success of this event, Emily threw her energy into helping other volunteer groups around the country organise anti badger cull protest marches. Working with badger groups, the police and local authorities, Emily soon became a key figure in the Badger Army.

Putting her energy and enthusiasm for badger protection to further good use, Emily has worked with the Badger Trust to develop National Badger Week into a highly successful campaign to raise awareness about badger behaviour and ecology and the need to protect the species.

Working with conservationists and broadcasters such as Chris Packham, Steve Backshall, Virginia McKenna and Bill Oddie to produce videos and support campaign events, Emily has helped bring the need for badger protection to a much wider audience.

Emily is in regular correspondence with Cameron, her constituency MP, on badger protection and other wildlife issues. When Liz Truss tried to avoid meeting the Badger Trust prior to the 2015 general election, it was Emily's intervention with the Prime Minister that helped secure the meeting.

Emily is a wonderful example of what can be achieved when anger is channelled into action.

DEBBIE BAILEY

Like many people, Debbie Bailey first became involved in campaigning for badgers as a result of them regularly coming into her garden, in Furness Vale in Derbyshire.

A nurse by profession, Debbie decided that the best way to prevent badger culling was to help farmers vaccinate badgers on their land against the spread of TB in the species.

Rather than simply donate to funds to an existing badger vaccination project, Debbie decided to train as a vaccinator and work with the High Peak Badger Group and the Derbyshire Wildlife Trust to expand the level of badger vaccination across Derbyshire.

Debbie has worked tirelessly with the farmers and landowners earning their trust and confidence to map badger setts and vaccinate badgers. She has become a key voice for the badger vaccination movement in Westminster and has helped raise public awareness about protecting badgers through her appearances in the print and broadcast media.

Nowhere else in the world will you find people like Debbie Bailey who are willing to give up their careers to go into the fields in all weathers to help protect wildlife. Debbie is a wonderful example of caring compassionate Britain.

NIGEL TOLLEY

Nigel has become a central figure in the badger protection movement both in the streets and the fields. As the Badger Army marched through towns and cities across the country, Nigel could be found at the front of the protest with a loud hailer in his hand shouting out: 'Save our badgers'.

A successful businessman and restaurant owner from the West Midlands, Nigel has played a major role in setting up a new Badger Trust Group in the West Midlands and has used his business skills to help develop the Birmingham Wildlife Festival into one of the most successful animal protection events in Europe.

Taking his camper van down into Gloucestershire for weeks at a time, Nigel has helped to organise and equip large numbers of volunteers who have played a vital role in the front line defence of badgers in the cull zones.

His laid-back nature and sense of humour hides a steely determination to stand up and protect wildlife. From rescuing injured badgers from the roadside to organising protest marches and wildlife festivals and protecting badgers in the cull zones, Nigel is one of the most effective wildlife activists in Britain today.

CHRIS WOOD

I first met Chris Wood in February 2014, when she organised a march against the badger cull in St Albans, involving her local Tory MP, Anne Main.

Chris plays a vital role in the operation of the Herts and Middlesex Badger Group, one of over 50 volunteer badger protection groups across the UK which are connected to the Badger Trust. These groups play a vital role in the rescue and rehabilitation of badgers, dealing with persecution incidents and raising awareness of badger ecology and behaviour.

Chris is a teacher by profession and has worked hard to bring the issue of badger and wider UK wildlife protection issues protection into the classroom. She is a tireless campaigner for badgers, rescuing injured animals, working closely with the police on persecution incidents, running fund-raising and badger awareness events, and joining marches and wounded badger patrols in the cull zones.

No matter the wildlife protection campaign issue or location of the protest, I can be certain that Chris and her husband Frank are on hand with their PA system and their beautiful little Serbian rescue dog Moby. In a nation of animal lovers, Chris stands out as an outstanding wildlife champion and a true friend of the badger.

SUSAN TIERNEY

During the 2015 election campaign, I travelled to Glasgow for a panel debate on wildlife protection issues. The organisers had hoped to involve both Labour and SNP MPs, but with Labour in political meltdown in Scotland, the Scottish Labour Party decided at the last minute to withdraw its support from the event. As result I made a six-hour train journey to Glasgow to find myself speaking to a largely empty hall. However, one person stood out in the audience with their energy and enthusiasm for badger and wildlife protection and that was Susan Tierney.

At the time Susan had gone from working in the social care sector to study zoology at university and pursue a career in wildlife conservation. At the time of our meeting, she was coming to an end of a contract with Scottish Badgers and was about to take up a new post with the League Against Cruel Sports in Scotland.

Following our meeting in Glasgow, I soon realised that Susan was a huge asset for the badger protection movement. Her expertise knowledge and enthusiasm impressed me greatly. Within six months of out meeting, Susan had relocated from Scotland to Sussex to take up a new post working alongside me to manage the Badger Trust.

Today Susan is helping reshape the Badger Trust to ensure it grows in strength not just as a campaigning and lobbying organisation, but also as an education and training organisation.

Susan has tireless energy a good sense of humour and a thick skin, all of which are vital in the badger protection world. In an NGO community which has become increasingly corporate and risk averse, Susan stands out as a highly committed wildlife defender who will roll up her sleeves to get the job done.

CLARE HAMMACOTT

On 21 February 2015, I spoke to thousands of people on my work campaigning for wildlife at the Birmingham Wildlife Festival in Centenary Square. Following the speeches, I joined a mass protest in the city against the badger cull. As we left Centenary Square, I was approached by Clare Hammacott, who told me she was working for the Commission on Civil Society on a review of the Lobbying Bill. Clare said she had been following my anti badger cull campaign activities and wanted to arrange a meeting to discuss how the Lobbying Bill could result in restrictions being placed on charities like the Badger Trust in terms of political lobbying and campaigning.

Following this first meeting, I worked closely with Clare to get across to politicians and policymakers the dangers of shutting down the voice of civil society for political objectives. As well as giving her time to support the Commission on Civil Society, Clare was also director of finance and operations at C40 Cities in London, a network of the world's largest cities set up to tackle climate change.

Clare had been a good friend of the charity sector and a champion of free speech and civic society, but she was also passionate about wildlife protection, particularly badgers.

In April 2015, Clare contacted me to say she had been diagnosed with terminal pancreatic cancer and had only a short time left to live. She asked me to meet with her to discuss a legacy for the Badger Trust.

Outside St Albans on a beautiful spring evening in late April, I met Clare at her sister's house. Her appearance told me all I needed to know about the ordeal she was going through. She had lost a significant amount of weight and was having to deal with the consequences of intense chemo-therapy, which at best would only prolong her life by months.

However, she had lost none of her sense of humour and she told me that facing death at the age of 42 made each day increasingly precious and she was determined to make the most of the limited time she had left. Despite her poor state of health, she had decided to make a trip to the Galapagos Islands with two close friends at the start of May.

Clare told me how pleased she was that the Badger Trust, unlike many larger charities, had been willing to push back at attempts by the government to stifle free speech and public opposition to its policies during the election campaign.

Clare was inspired by the Badger Army movement and the way it had mobilised thousands of people across the country to come together in defence of the badger. She made it clear to me that she would not leave a legacy to any of the large environmental NGOs, instead she wanted to do all she could to help use some of her estate to support the Badger Trust beyond her death.

Meeting Clare in such difficult circumstances was both a very difficult but also uplifting moment. I can only hope that when facing my own death, I can show the same courage and determination to make the most of each remaining day.

Clare's courage and commitment to protect badgers beyond her death filled me with more determination than ever to fight on against the incompetence, negligence and deceit behind the badger cull.

This book is dedicated to Clare Hammacott
and all those people who despite their own pain and suffering
remain dedicated to protecting our precious wildlife.

INDEX

C

Canbury Press

Telling the real story since 2013

Canbury Press publishes high-quality non-fiction revealing the underlying truth about modern life.

These new titles are available at good bookshops and Canburypress.com (free post and packing in UK):

— *Order today and start reading this week*

'An intense heart-rending roller-coaster of a book'
Will Black, author

BECAUSE WE ARE BAD
OCD and a Girl
Lost in Thought
Lily Bailey

256 pages
Hardback (£14.99)
ISBN: 9780993040726
Ebook (£6.99)
ISBN: 9780993040733

A highly-acclaimed, sad, and amusing account of a childhood wracked by Obsessive Compulsive Disorder

The Witch in the Broom Cupboard

AND OTHER TALES

Pushkin Children's Books
71–75 Shelton Street
London, WC2H 9JQ

Original Text © Éditions de la Table Ronde, Paris, 1967
Illustrations by Puig Rosado
Translation © Sophie Lewis 2013

This edition published by Pushkin Children's Books in 2014

These translations were first published by Pushkin Children's
Books as part of *The Good Little Devil and Other Tales* in 2013

0 0 1

ISBN 978-1-78269-066-5

This book is supported by the Institut français
Royaume-Uni as part of the Burgess programme
(www.frenchbooknews.com)

Set in 12 on 19 Berling Nova by Tetragon, London

Printed in China by WKT Co

www.pushkinpress.com

The
Witch in the
Broom Cupboard
AND OTHER TALES

by Pierre Gripari

Illustrated by Puig Rosado

Translated from the French by Sophie Lewis

PUSHKIN CHILDREN'S BOOKS

Contents

The Witch of Rue Mouffetard

There was once an old witch living in the Gobelins neighbourhood in Paris; she was a dreadfully old and ugly witch, but she really did want to be the most beautiful girl in the world!

One sunny day, while reading the *Witches' Times*, she came across the following advertisement:

MADAME
You who are OLD and UGLY
You shall become YOUNG and PRETTY!
To achieve this:
EAT A LITTLE GIRL
In tomato sauce!

Underneath, in small letters, it said:

BUT TAKE CARE!
YOUR LITTLE GIRL'S FIRST NAME
ABSOLUTELY MUST BEGIN
WITH THE LETTER N!

Now, a little girl whose name was Nadia happened to be living in the very same neighbourhood as the witch. She was the eldest daughter of Papa Sayeed (perhaps you know him?), who kept the cafe-grocer's on rue Broca.

"I shall have to eat Nadia," the witch decided.

One fine day, Nadia had gone out to get some bread from the bakery when an old lady stopped and spoke to her:

"Good morning, young Nadia!"

"Good morning, madame!"

"Would you like to do me a good turn?"

"What is it?"

"Would you go and fetch a tin of tomato sauce from your daddy's shop for me? It would save me going, and I'm so tired today!"

Nadia agreed right away; she was a good-hearted

girl. As soon as she had gone, the witch—for it was she—began to laugh and rub her hands together:

"Oh, I am so cunning!" she said. "Young Nadia is going to bring me the very sauce that I shall eat her with!"

As soon as she had come back home with the bread, Nadia took a tin of tomato sauce from the shelves, and she was just getting ready to go out again when her father stopped her:

"And where are you off to, with that?"

"I am taking this tin of tomato sauce to an old lady who asked me for it."

"You stay here," said Papa Sayeed. "If your old lady wants something, she has only to come here herself."

Nadia, being also a very obedient girl, did not argue. But the next day, while out shopping, she was stopped by the old lady once again:

"Well, Nadia? What about my tomato sauce?"

"Sorry," said Nadia, blushing from head to foot, "but my daddy didn't want me to bring it. He said you should come to the shop yourself."

"Very well," said the old lady, "I'll come, then."

Indeed, she walked into the shop that very same day:

"Good morning, Monsieur Sayeed."

"Good morning, madame. What can I get you?"

"I would like Nadia."

"Excuse me?"

"Oh, forgive me! I meant to say: a tin of tomato sauce, please."

"Of course! A small one or a large one?"

"A large one, it's for Nadia..."

"What?"

"No, no! I meant to say: it's to have with some spaghetti..."

"I see! Talking of which, we also have spaghetti..."

"Oh, there's no need, I'll have Nadia..."

"What?"

"Do forgive me! I meant to say: the spaghetti, I already have some at home..."

"If you're sure... Here is your tomato sauce."

The old lady took the tin and paid for it, but then, instead of leaving, she began to look doubtful:

"Hm! Perhaps it is a little heavy... Do you think you might perhaps..."

"Might what?"

"Let Nadia carry it home for me?"

But Papa Sayeed had his suspicions.

13

"No, madame, we don't deliver. Besides, Nadia has other things to be getting on with. If this tin is too heavy for you, well, too bad: you'll just have to leave it behind!"

"Never mind," said the witch, "I'll take it. Goodbye, Monsieur Sayeed!"

"Goodbye, madame!"

And the witch went away, with her tin of tomato sauce. As soon as she was home, she said to herself:

"Here's an idea: tomorrow morning, I shall disguise myself as a butcher, then go to rue Mouffetard and sell some meat in the market. When Nadia comes out to do the shopping for her parents, I'll nab her."

The following day, the witch appeared on rue Mouffetard disguised as a market butcher, when Nadia happened to go by.

"Hello, little girl. Would you like some meat?"

"Oh no, madame, I've just bought a chicken."

"Shoot!" thought the witch.

Next day, back in the market, she had disguised herself as a poultry butcher.

"Hello, dear. Will you buy one of my chickens?"

14

"Oh no, madame. Today I'm looking for some red meat."

"Blast!" thought the witch.

On the third day, in a fresh disguise, she was selling both red meat and poultry.

"Hello Nadia, hello my dear! What would you like? You see, today I have something for everyone: beef, mutton, chicken, rabbit..."

"Yes, but we're having fish today!"

"Drat!"

Back at home, the witch thought and thought. Then she had a new idea:

"Well, if this is how things are, I will use some stronger magic. Tomorrow morning I shall turn myself into EVERY SINGLE ONE of the food-sellers on rue Mouffetard AT THE SAME TIME!"

And indeed, the following day, the witch had turned into every single one of the food-sellers on rue Mouffetard (there were exactly 267 of them), in disguise.

Nadia came along as usual and, quite unsuspecting, went up to a vegetable stall—to buy some green beans, this time—and was about to pay when the shopkeeper caught her by the wrist, snatched her away and *ker-CHING!* shut her up in the till.

Luckily, Nadia had a little brother whose name was Bashir. Noticing that his big sister had not come home, Bashir said to himself:

"That witch must have taken her. I have to go and save her."

He picked up his guitar and headed off to rue Mouffetard. Seeing him approach, the 267 food-sellers (remember: every single one of them was actually the witch) began to call out:

"Where are you off to like that, Bashir?"

Bashir closed his eyes tight and answered:

"I am a poor blind musician; all I want is to sing a little song and earn myself a few pennies."

"What song?" the food-sellers asked.

"I want to sing a song called: *Nadia, Where Are You?*"

"No, not that one! Sing another!"

"But I don't know another song!"

"Then sing it really softly!"

"All right! I'll sing it really softly."

And Bashir began to sing as loudly as he could:

> *Nadia, where are you?*
> *Nadia, where are you?*
> *Reply so I can spy you!*
> *Nadia, where are you?*
> *Nadia, where are you?*
> *You've vanished from view!*

"Softer! Softer!" cried the 267 food-sellers. "You're hurting our ears!"

But Bashir went on singing:

> *Nadia, where are you?*
> *Nadia, where are you?*

When suddenly a little voice replied:

> *Bashir, Bashir, set me free*
> *Or the witch will kill me!*

At these words, Bashir opened his eyes and all of the 267 food-sellers leapt upon him, screeching:

"He's faking! He's faking! He can see!"

But Bashir, who was a brave boy, swung his small guitar and knocked over the nearest stallholder with a single blow. She fell flat on the ground, and the other 266 fell over all at once too, stunned just like their colleague.

Now Bashir went into all the shops on the street, one after the other, singing:

Nadia, where are you?
Nadia, where are you?

Once more, the little voice replied:

Bashir, Bashir, set me free
Or the witch will kill me!

This time there was no doubt: the voice was coming from the grocer's shop. Bashir raced inside, leaping over the vegetable display, just as, coming round from her faint, the witch-grocer opened her eyes. And, just as she came to, the other 266 food-sellers also opened their eyes. Luckily, Bashir saw her in time and, with a well-aimed blow from his guitar, he knocked them all out again for a few minutes longer.

Then, he tried to open the till, while Nadia continued to sing:

Bashir, Bashir, set me free
Or the witch will kill me!

But the drawer was too tightly closed; it wouldn't move an inch. Nadia was singing and Bashir was struggling, and all the while the 267 witch-food-sellers were waking up again. But this time, they took good care not to start opening their eyes! Instead, they kept their eyes closed and began to crawl towards the grocer's where Bashir was working away, so as to surround him.

Just then, when exhausted Bashir couldn't think which way to turn next, he saw a tall sailor go past, a well-built young man, walking down the street.

"Hello, sailor. Would you mind helping me out?"

"What can I do?"

"Could you carry this shop's till all the way to our house? My sister is stuck inside it."

"And what will my reward be?"

"You shall have the money and I'll have my sister."

"It's a deal!"

Bashir lifted the till and was just about to pass it over to the sailor when the witch-grocer, who had crept up quietly as a mouse, caught him by the foot and began to squeal:

"Ah, you thief, I have you now!"

Bashir lost his balance and dropped the till. Being very heavy indeed, when the till fell straight onto the witch-grocer's head, the single blow cracked open the heads of all 267 witch-food-sellers and knocked their brains out. This time the witch was dead, well and truly dead.

And that's not all: with the force of the impact, the till drawer flew open—*ker-CHING!* And Nadia stepped out.

She hugged and thanked her little brother, and the pair of them went home to their parents, while the sailor gathered up all the witch's blood-spattered money.

The Giant Who Wore Red Socks

There was once a giant who always wore bright-red socks. He was three storeys tall and lived underground.

One fine day, he said to himself:

"It's boring to stay a bachelor! Let me take a look around up there and see if I can get myself a wife."

No sooner said than done: he knocked a big hole in the ground above his head... but unfortunately, instead of popping up out among meadows, he ended up in the middle of a village.

In this village there was a young girl whose name was Mireille and who loved eating soft-boiled eggs. That particular morning, she was in fact sitting down with an egg in its egg cup, getting ready to crack it open with a teaspoon.

At the first tap of the spoon, the house began to shake.

"Gosh! Have I suddenly got stronger?" Mireille wondered.

At the second tap of the spoon, the house began to move.

"If I go on like this," she thought, "I shall bring the house right down. Perhaps it would be better if I stopped."

But since she was hungry, and she really did love soft-boiled eggs, she decided to go on all the same.

At the third tap that Mireille gave her egg, the whole house flew into the air, like a champagne cork, and, in its place, poking out of the ground, appeared the giant's head.

The young lady was herself thrown into the air. Luckily she landed in the giant's hair, so she wasn't at all hurt.

But now, running his fingers through his hair in order to shake the rubble out, the giant felt her wriggling there:

"Goodness!" he thought. "What have I got in there? Feels like some kind of creature!"

He pulled the creature out and peered at it:

"What are you?"

"I am a girl."

"What is your name?"

"Mireille."

"Mireille, I love you. I want to marry you."

"First put me down, and then I'll give my answer."

The giant put her back on the ground and Mireille ran away as fast as her legs could carry her, screaming: "Aaaaaaaaaah!"

"What did she mean by that?" wondered the giant. "That's not an answer!"

All the same, he finished pulling himself out of the ground. He was just straightening his trousers when the village mayor and vicar came along. They were both very angry.

"What on earth is this? A fine way to go about your business! Popping out of the ground like this, plumb in the middle of a residential area... Where exactly do you think you are?"

"I do apologize," replied the giant, "I didn't do it on purpose, I promise."

"And poor Mireille!" exclaimed the vicar. "Her house is quite ruined!"

"If that's all," said the giant, "then it isn't so terrible. I'll rebuild it myself!"

And there and then, he spoke the following magic words:

"By the power of my bright-red socks, let Mireille's house be set aright!"

Instantly, the house became just as it had been before, with all its walls, doors, windows and furniture, its dusty corners and even its spiderwebs! Mireille's soft-boiled egg was back in its egg cup, piping hot all over again, ready for her to eat it!

"That's better," said the vicar, calming down. "I see you're not bad at heart. Now, be on your way."

"One moment, please," said the giant. "I want to ask you something."

"What now?"

"I would like to marry Mireille."

"That's impossible," the vicar replied.

"Impossible? Why?"

"Because you are too tall. You will never fit inside the church."

"It is true that the church is very small," said the giant. "What if I blow some air inside to make it a little bigger?"

"That would be cheating," said the vicar. "The church must stay as it is. It is you who must shrink."

24

"I would like nothing better! How should I go about shrinking?"

There was a silence. The mayor and the vicar exchanged looks.

"Listen," said the vicar, "I like you. Let me send you to see the great Chinese wizard. While you are away, I will speak to Mireille. Come back in one year and she will be ready to marry you. But take care! She will not wait longer than a year!"

"And where does your Chinese wizard live?"

"In China."

"Thank you."

And the giant set off. It took him three months to reach China and another three months to find the wizard. He spent this time learning to speak Chinese. Standing, at last, before the wizard's house, he knocked at the door. The wizard answered the door and the giant said to him:

"Yong cho-cho-cho kong kong ngo."

Which in Chinese means: "Are *you* the great wizard?" To which the wizard replied, in a slightly different tone:

"Yong cho-cho-cho kong kong ngo."

Which means: "Yes, it's me. So?"

(Chinese is like that: you can say almost everything with a single sentence, as long as you change the intonation.)

"I would like to be shrunk," said the giant, still in Chinese.

"Fine," said the Chinese wizard, also in Chinese, "wait a minute."

He went inside, then came back with a cup full of magic potion. But the cup was too small: the giant couldn't even see it. So the wizard vanished inside again and came back with a bottle. But the bottle was too small: the giant couldn't even pick it up.

Then the wizard had an idea. He rolled his big barrel of magic potion out of the front door, then set it upright and opened it up at the top. The giant drank from the barrel just as we drink from a glass.

When he had finished drinking, he waited. Now, not only did he stay the same size but, from being bright red before, his socks turned green. The great Chinese wizard had simply given him the wrong magic potion.

Then the giant got very angry and yelled very loudly:

"Yong cho-cho-cho kong kong ngo!"

Which means: "Are you trying to make a fool of me?"

The wizard apologized and came back with another barrel, which the giant drank and his socks went red again, as they had been before.

"Now, shrink me," said the giant to the Chinese wizard, still in Chinese.

"I do apologize," said the wizard, "but I've run out of potion."

"Now what am I going to do?" cried the giant, in a desperate tone.

"Listen," said the Chinese wizard, "I like you. Let me send you to see the great Breton wizard."

"And where does your Breton wizard live?"

"In Brittany."

So the giant went on his way, saying:

"Yong cho-cho-cho kong kong ngo."

Which means "Thank you!" And the Chinese wizard watched him go, calling after him:

"Yong cho-cho-cho kong kong ngo!"

Which means: "My pleasure. *Bon voyage!*"

Three months later, the giant landed in Brittany. It took him yet another month to find the Breton wizard.

"What do you want?" asked the Breton wizard.

The giant replied:

"Yong cho-cho-cho kong kong ngo."

"Pardon?"

"Forgive me," said the giant, "I thought I was still in China. I meant to say: could you make me smaller?"

"That's very easy," said the Breton wizard.

He went into his house, then came out again with a barrel of magic potion.

"Here, drink this."

The giant drank it but, instead of shrinking, he began to grow, and was very soon twice as tall as before.

"Oh I am sorry!" said the wizard. "I must have picked up the wrong barrel of potion. Just stay there, I won't be a second."

He disappeared, and came back with another barrel.

"Here, drink this one," he said.

The giant drank and... so it proved. He shrank back to his usual height.

"This is not enough," he said. "I need to be as small as a man."

"Ah, that small? I'm sorry, that's not possible," said the wizard, "I'm all out of potion. Come back in six months."

"But I can't!" exclaimed the giant. "I must return to my fiancée within the next two months!"

And, saying this, he began to cry.

"Listen," said the wizard, "I like you, and besides, this is my fault. In view of this, I shall give you a good recommendation. Let me send you to see the Pope of Rome."

"And where does he live, this Pope of Rome?"

"In Rome."

"Thank you very much."

One month later, the giant arrived in Rome. It took him another fortnight to find the Pope's house. Once he had found it, he rang the doorbell. After a few seconds, the Pope came to the door.

"Sir... What can I do for you?"

"I want," the giant said, "to become as small as a man."

"But I am not a wizard!"

"Have pity on me, Mr Pope! My fiancée is expecting me in a fortnight!"

"What then?"

"Well, if I'm still too tall then, I won't be able to get inside the church in order to marry her!"

Hearing this, the Pope felt sorry for the giant:

"That would be sad!" he said. "Listen, my friend, I like you. I shall try to do something for you."

The Pope went into his house, picked up the telephone and dialled these three letters: HVM.

Perhaps you know, when you dial O, you are put through to the Operator. But what you may not know is that when you dial HVM, you come through to the Holy Virgin Mary. If you don't believe me, wait for a day when your parents are out, and try it!

Indeed, after a few moments, a gentle voice could be heard:

"Hello? Holy Virgin here. Who is speaking?"

"It's me, the Pope!"

"You? Ah, how lovely! And what do you want?"

"Well, it's like this: I have a giant here, who would like to become as small as a man. In order to get married, as far as I understand..."

"And does this giant of yours wear bright-red socks with special powers?"

"So he does, Holy Virgin! How did you know?"

"Well, you see, I just know!"

"Really, Holy Virgin, you are a marvel!"

"Thank you, thank you... Now, tell your giant that he should leave his red socks at the laundrette and go and soak both his feet in the sea, while calling my name. He shall see what happens next!"

"Thank you, Holy Virgin."

"That's not all! As I predict that he will still have a few problems, tell him that, afterwards, he can have three wishes, which will come true straight away. But he must be careful! Three wishes, no more!"

"I will tell him."

And the Pope repeated to the giant what the Holy Virgin Mary had told him.

Later that day, the giant handed his red socks in at the laundrette, then he went to the very edge of the sea, paddled his bare feet in the blue water, and began to call out:

"Mary! Mary! Mary!"

Pouf! He lost his footing and fell over straight away. He had become as small as a man. He swam back to shore, dried himself in the sun and went back to the laundrette.

"Good morning, madam. I've come to pick up my red socks."

"I don't have any red socks here."

"But you do! The pair of red socks, about three metres long..."

"Ah you mean: the two red sleeping bags?"

"They're socks I tell you!"

"Listen," said the laundrette assistant, "call them what you will, but when I see a sock that I can lie down in, I call it a sleeping bag!"

"Never mind, please give them back to me!"

But when he tried to put his red socks on, the poor man realized that they now came up above his head. He began to cry:

"What is to become of me? I am no longer a giant and, without my magic red socks, I'm nobody! If only they too could be shrunk down to my size!"

No sooner had he said this than his red socks shrank too, and he was able to put them on. His first wish was granted.

Very happy, he put his shoes back on and thanked the Holy Virgin, after which he thought about going back to the village where he had started out.

However, since he was no longer a giant, he could not walk all the way back to Mireille's village. Moreover, he didn't have the money to take a train. Once more he burst into tears:

"Alas! And I've only got a fortnight to get back to my fiancée! If only I could be near her!"

No sooner had he said this than he found himself in Mireille's dining room, just as the young lady herself was about to crack open a soft-boiled egg. As soon as she saw him, she jumped up and threw her arms around him:

"The vicar explained everything," she said. "I know all about what you have done for me, and now I am in love with you. In six months' time, we shall be married."

"Only in six months' time?" asked the man in red socks.

But then he had a sudden thought—that he still had his third wish to make, and so he said aloud:

"Let today be our wedding day!"

No sooner had he said this than he was stepping out of the church, in bright-red socks and a fine black suit, with Mireille at his side, all dressed in white.

From that day onwards, they lived very happily together. They have many children and the former giant, their father, earns enough for the whole family by building houses, which is easy for him, thanks to the magical powers of his bright-red socks.

Scoobidoo, the Doll
Who Could See Everything

There was once a little boy whose name was Bashir. He had a rubber doll called Scoobidoo and a papa called Sayeed.

Sayeed was a good papa, just like the good papas we all know, but Scoobidoo was no ordinary doll: she had magical powers. She walked and talked just like a person. What's more, she could see into the past and the future, and she could see things that were hidden. For her to do that, all they had to do was put a blindfold over her eyes.

She often played dominos with Bashir. When she had her eyes open, she always lost, as Bashir played better than she did. But when he blindfolded her, it was Scoobidoo who won.

One fine morning, Bashir said to his father:

"Papa, can I have a bicycle?"

"I don't have enough money," said Papa Sayeed. "Also, if I buy you a bicycle now, next year you will have grown and it will be too small. Later, in a year or two, we can think about it again."

Bashir did not mention it again, but that evening he asked Scoobidoo:

"So tell me, since you can see everything: when will I get a bicycle?"

"Blindfold me," said Scoobidoo, "and I'll tell you."

Bashir took a cloth and tied it over her eyes. Straight away Scoobidoo said:

"Yes, I do see a bicycle... But it's not for right now... It's in a year or two..."

"No sooner?"

"No sooner!"

"But I want one right now!" shouted Bashir angrily. "Look: you have magic powers, don't you?"

"I do," agreed Scoobidoo.

"Then make Papa buy me a bicycle."

"I would be happy to try, but it won't work."

"Never mind! Try anyway."

"All right: leave me blindfolded overnight, and I will try."

That night, while everyone was sleeping—Papa, Mama, Bashir and his elder sisters—Scoobidoo, in her corner, began singing very softly:

> *Papa wants a bicycle*
> *A very little bike*
> *Swift as a kite*
> *With two wheels*
> *Silent as seals*
> *With a saddle*
> *Solid as cattle*
> *With brakes*
> *Canny as snakes*
> *With a headlight*
> *Bright as a sprite*
> *And a bell*
> *Sound as a gazelle*
> *It's for Bashir*
> *Swift as a deer!*

All through the night, she sang this magical song. She stopped singing at dawn, for the magic was complete.

That morning, Papa Sayeed went to do some shopping on rue Mouffetard. To start with, he went to the baker:

"Good morning, madame."

"Good morning, Papa Sayeed. What would you like?"

"I would like a bicycle," said Papa Sayeed.

"What did you say?"

"Goodness, what am I saying? I mean: a two-pound loaf, please."

Next, Papa Sayeed went to the butcher.

"Hello, Papa Sayeed. What will it be today?"

"A good one-and-a-half pound of bike," said Papa Sayeed.

"Ah, I'm very sorry," said the butcher. "I have beef, mutton and veal, but I don't sell bicycles."

"But what nonsense am I coming out with? Of course! I meant to say: a good joint of beef, please!"

Papa Sayeed took the joint, paid and went to the grocer's next.

"Hello, Papa Sayeed. What can I get for you?"

"A pound of nice ripe bicycles," said Papa Sayeed.

"A pound of what?" asked the grocer.

"What is wrong with me, today? A pound of ripe white grapes, please!"

This is how it was, all day. Every time Papa Sayeed went into a shop, he began by asking for some bicycle. Just like that, without meaning to; he couldn't help it. So it happened that he asked for another box of white bicycles in the grocer's, a good slice of bicycle at the cheese shop, and a bottle of bicycle bleach at the laundrette. Finally, very worried, he dropped in to see his doctor.

"Hullo there, Papa Sayeed, what is the problem?"

"Well, it's like this," said Papa Sayeed. "Since this morning, I don't know what's wrong with me but, each time I go into a shop, I start by asking for a bicycle. It's against my will, I assure you, I'm not doing it on purpose in the least! What is this sickness? I'm very disturbed by it... Couldn't you give me a little bicycle... There you are! It's happening again! I mean a little medicine, to stop it happening?"

"Ahem," said the doctor. "Curious, very curious indeed... Tell me, Papa Sayeed, you wouldn't happen to have a young son, by any chance?"

"Yes, I do, doctor."

"And this young son has recently asked you for a bicycle..."

"How do you know that?"

"Heh heh! It's my job! And your son would not happen to have a doll, by any chance? A rubber doll known as Scoobidoo?"

"He does indeed, doctor!"

"So I thought! Well then, watch out for that doll, Papa Sayeed. If she is to stay with you, she will make you buy a bicycle, whether you like it or not. And—that will be three thousand francs!"

"Oh no! They're much more expensive than that!"

"I'm not talking about a bicycle, I'm talking about your medical consultation. You owe me three thousand francs."

"Yes, of course!"

Papa Sayeed paid the doctor, went back home and said to little Bashir:

"Would you do me a favour and get rid of your doll, because if I find her I shall throw her in the fire!"

As soon as Bashir and Scoobidoo were alone:

"You see," said Scoobidoo, "I did tell you that it wouldn't work... But don't be sad. I will go away and

I'll come back in a year's time. On my return, you shall have your bicycle. However, there's something I will need before I go..."

"What's that?" asked Bashir.

"Well, when I'm alone, you will no longer be there to blindfold me... So I would like you to make me a pair of glasses with wooden lenses."

"But I don't know how to do that!"

"Ask your Papa."

Only too happy to know the doll was leaving, Papa Sayeed agreed to make her a pair of glasses with wooden lenses. He cut the lenses out of a piece of plywood with a little saw, made a frame out of wire, and said to Scoobidoo:

"How do you like these?"

Scoobidoo tried on the glasses. They suited her straight away.

"Very nice, sir, thank you very much."

"Okay. Now, out you go!" said Papa Sayeed.

"Of course. Goodbye, sir. Goodbye, Bashir."

And Scoobidoo left.

She journeyed for a long, long time, walking by night and hiding by day, in order not to attract attention. After three weeks, she came to a great

port on the shore of the English Channel. It was night-time. A large ship was at the dock, ready to depart early the following morning and begin its long journey around the world.

Putting on her glasses, Scoobidoo thought to herself:

"This ship is just right for me."

She put her glasses back in her pocket, then stationed herself at the foot of the gangway, and there she waited.

On the stroke of three in the morning, a sailor walking in zigzags rolled up to the gangway and was about to step onto it when he heard a tiny voice calling to him from ground level:

"Mister sailor! Mister sailor!"

"Who's there?" asked the sailor.

"Me, Scoobidoo! I'm right in front of you. Look out, you're about to step on me!"

The sailor crouched down:

"Well I never! How strange. A talking doll! And what might you be after?"

"I want you to take me onto the ship with you."

"And how will you make yourself useful?"

"I can see into the future and predict the weather."

"Really! Then, tell me what the weather will be tomorrow morning."

"Just a moment, please."

Scoobidoo took out her glasses, put them on, then said without hesitation:

"Tomorrow morning the weather will be bad. So bad that you won't be able to leave the port."

The sailor burst out laughing.

"Ha ha ha! You know nothing about it! We'll have fine weather, as it happens, and we weigh anchor at dawn."

"And I say you won't be able to leave!"

"All right then, let's bet on it, if you like? If the weather is good enough for us to go, I will leave you behind. But if bad weather holds us back, I'll take you along. Is it a deal?"

"It's a deal."

And so it happened that, the following day, the sun had hardly risen when a great cloud appeared in the north-west and spread so, so quickly that within five minutes the whole sky had turned black. Then the storm broke, so wild and violent that the ship was obliged to stay where she was.

"I don't understand at all," said the captain. "The forecast said it would be fine!"

"Well," said the sailor, "I know a doll that predicted this bad weather."

"A doll? You'd had a few drinks, hadn't you?"

"I'd drunk a good many," said the sailor, "but even so. She's a little rubber doll called Scoobidoo."

"And where is this Scoobidoo?"

"There, on the dock; I can see her from here."

"Bring her here."

The sailor leant over the rail and called:

"Here, Scoobidoo! Come up, will you? The captain would like to talk to you."

As soon as Scoobidoo was on board, the captain asked her:

"What is it that you can do, exactly?"

And Scoobidoo replied:

"I can see into the past and the future, and I can see things that are hidden."

"Is that all you can do? All right, tell me something about my family!"

"Right away, captain!"

Putting on her glasses, Scoobidoo began to speak very quickly, as if she were reading from a book:

"You have a wife in Le Havre, with a blond child. You have a wife in Singapore, with two yellow children. You have a wife in Dakar, with six black children..."

"Enough, enough!" exclaimed the captain. "You can come with me. Don't say another word!"

"And how much will you pay me?" asked Scoobidoo.

"Well, how much do you want?"

"I would like five new francs per day, to buy a bicycle for Bashir."

"Done. You'll be paid on your return."

So it was that Scoobidoo set off to sail around the world. The captain tied her to the bulkhead in his cabin with a pink ribbon, and every morning he asked her:

"Will it be rain or shine today?"

Thanks to her glasses, Scoobidoo replied correctly every time.

The great ship sailed right around Spain, on past Italy, past Egypt, through the Indian Ocean, past Thailand, through the Pacific Ocean, crossed into the Atlantic via the Panama Canal, then headed back towards Europe.

*

One fine morning, when the ship was nearing the coast of France, the cook snuck into the captain's cabin. Scoobidoo enquired:

"What are you doing in here?"

"Have a guess," replied the cook.

Scoobidoo took out her glasses, put them on and gasped:

"You've come to steal my glasses!"

"Right you are," said the cook.

And, before Scoobidoo had time to think, he snatched them from her, slipped back out of the cabin and threw them into the sea.

A few minutes later, as if on cue, the captain came back into the cabin.

"So tell me, Scoobidoo, what weather shall we have tomorrow morning?"

"I can't say," replied the doll, "the cook has stolen my glasses."

The captain raised an eyebrow: "Glasses or no glasses, you promised to forecast the weather. What do you think? That I'll keep on paying you for nothing?"

The captain was pretending to be angry, but in fact it was he who had sent the cook to steal the glasses, because he did not want to pay Scoobidoo what he owed her.

"You may get along however you like," he said, "but if you don't tell me what weather we'll have tomorrow morning, I shall throw you into the sea!"

"All right... let's say: it will be sunny!" said Scoobidoo, making a guess.

Alas! As early as sunrise the next day, a fat black cloud appeared on the horizon and began to spread rapidly, as if it was trying to gobble up the sky. At the same time, a storm was setting in and the ship began to pitch from side to side. The captain came in—or pretended to come in—in a terrible fury.

"You have deceived me!" he thundered at Scoobidoo.

And then, paying no attention to her protests, he threw her overboard.

Poor, dazed Scoobidoo saw the sea and the sky spin around her before she dropped into the water. Almost immediately, a great mouth full of pointed teeth opened wide beneath her, and she was swallowed up by a shark that had been following the ship for several days.

Since the shark was very greedy, it swallowed her without chewing, so that Scoobidoo found herself in its stomach, not too comfortable, but not the least bit

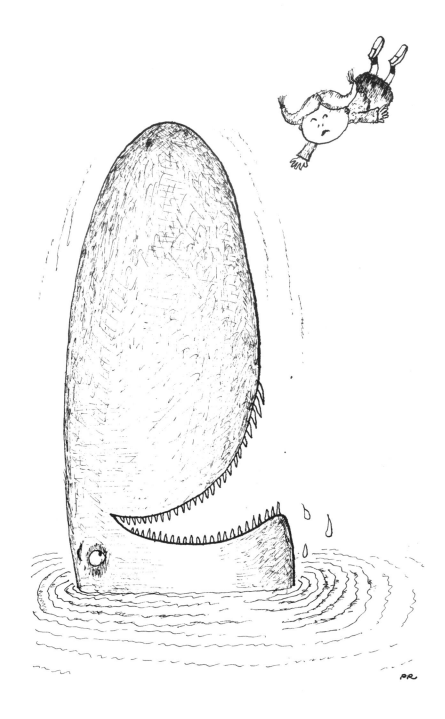

hurt. She tried to feel her way about in the dark, all the while muttering to herself:

"What will become of me here? And what about my poor little Bashir, still waiting for his bicycle?"

So it was that, talking aloud in the pitch darkness, Scoobidoo came across something that felt like a miniature bicycle: it had two round blocks of wood, linked together by a wire frame.

"Well I never... my glasses!"

They were indeed her glasses, which the shark had swallowed the day before. Scoobidoo picked them up, put them on and straight away saw everything as clear as day, there inside the stomach of the great fish. She exclaimed happily:

"And there's treasure in here!"

Upon which, without the least hesitation, she turned towards a fat oyster that was lying, wide open, in a fold in the shark's stomach.

"Hello, oyster!"

"Hello, doll!"

"Am I right in thinking you have a big pearl inside you there?"

"Alas, you are quite right!" replied the oyster, sighing. "A very big pearl, which is hurting me horribly! If only I could find someone to take this lump of dirt away!"

"Would you like me to take it?"

"Now, if you were to do that, you would be doing me a great service!"

"Open yourself up very wide, then, and we shall see!"

The oyster opened up as wide as she could. Scoobidoo plunged both hands in and plucked out the pearl.

"Ooowww!" exclaimed the oyster.

"There, there, now, it's all over."

And Scoobidoo held up the pearl. It was enormous, a magnificent pearl. It was worth enough money to buy five or six bicycles! Scoobidoo put it into her pocket and said politely to the oyster:

"Thank you."

"Not at all—thank *you*! If I can do anything for you..."

"You might be able to give me some advice," said Scoobidoo.

"Of course!"

"What should I do to make my way home?"

"It's very simple," said the oyster. "Since you have two legs, you have only to hop from one foot to the other. That will make the shark feel sick, and he will do anything you ask."

"Thank you, kind oyster!"

And Scoobidoo began hopping from one foot to the other.

After a minute, the shark began to feel unwell. After two minutes, he had hiccups. After three minutes, he was seasick. After five minutes, he called out:

"Hey, are you quite finished in there? Can't you sit quietly and let me digest you?"

"Take me to Paris!" Scoobidoo called to him.

"To Paris? Whatever next? I do not take orders from my food!"

"In that case, I'll keep hopping!"

"No, no! Stop! Where is it, this Paris?"

"You get there by swimming up the River Seine."

"Err—what? Swim up the River Seine? But I shall be a laughing stock! I am a fish of the open ocean! No one in my family has ever left salt water!"

"Then I'll keep on hopping!"

"No, no! Have pity! I'll go wherever you wish. But do stay still a bit!"

And the great fish set off. He swam as far as the port of Le Havre, then he swam up the River Seine, through Rouen and all the way to Paris. Once there, he stopped beside a stone staircase at the water's edge, opened his mouth and called out as loudly as he could:

"End of the line; all change please! All change! Out you get, now, scarper. I hope I never see you again!"

Scoobidoo got out and climbed up onto the riverbank. It was about three o'clock in the morning. No one was in the streets, nor was there even one star in the sky. Taking advantage of the darkness and with the help of her glasses, the doll quickly made her way back to rue Broca. The following morning, she knocked on Papa Sayeed's door and handed him the pearl. Papa Sayeed thanked her, then took the pearl to the jeweller, and was at last able to buy a bicycle for Bashir.

As for the ship on which Scoobidoo had sailed away, it was never seen again. I do believe it came to a watery end.

The Witch in the Broom Cupboard

It's Monsieur Pierre writing here, and now I'm going to tell you a story about something that really happened to me.

Rummaging in my pocket one day, I found a five-franc coin. I thought to myself:

"Hurray! I'm rich! Now I can buy myself a house!"

And I hurried off to see a solicitor:

"Good morning, monsieur. Would you happen to have a house to sell for about five francs?"

"Five francs? No, I'm terribly sorry," said the solicitor, "I haven't got one for that price. I have houses at twenty francs, at fifty francs, at a hundred francs, but nothing at five francs."

I asked again, just to be sure:

"Really? Might we not find one if we looked very hard... Not even a very small house?"

And then, the solicitor slapped his forehead:

"Now you mention it, I may have something! Just a moment..."

He rummaged through his files and pulled out a folder:

"Look, here you go: a neat little house on the high street, with one bedroom, kitchen, bathroom, living room, toilet and broom cupboard. "

"How much is it?"

"Three francs fifty. With my fees on top, it will come to five francs exactly."

"That's perfect. I'll take it."

Proudly, I laid my five-franc coin down on the desk. The solicitor took it and held out a contract.

"Here you go, please sign here. And put your initials there. And there. And there too."

I signed everywhere and handed back the contract, saying:

"Is this all right?"

He replied:

"Quite right. He he he he!"

I stared at him, intrigued:

"What are you laughing for?"

"Oh, nothing, nothing... Haha!"

I didn't much like that laugh. It was a nervous little laugh, the laugh of someone who has just played a mean trick on you. I asked again:

"This house does exist, doesn't it?"

"Absolutely! Heh heh heh!"

"And is it well built? The roof isn't about to come down on my head, is it?"

"Hoho! Certainly not!"

"In that case, what's so funny?"

"Nothing at all, like I said! Anyway, here; have the key. You can go and see for yourself... Good luck! Hoo hoo hoo!"

I took the key and left to go and see the house. It was indeed a very pretty little house, smartly fitted, bright and airy, with a bedroom, kitchen, bathroom, living room, toilet and broom cupboard. When I had looked into every room, I said to myself:

"What about a quick hello to my new neighbours?"

Okay, let's get going! I went to knock on my neighbour's door to the left.

"Hello neighbour! I'm your new neighbour on the right. I've just bought the little house with a bedroom, kitchen, bathroom, living room, toilet and broom cupboard."

Hearing this, the good man turned as white as a sheet, before my eyes. He looked at me with a horrified expression, and *bam*! Without a word, he slammed the door in my face. I thought to myself, charitably:

"Well! Quite an eccentric!"

And I went to knock on the door of my right-hand neighbour:

"Hello neighbour! I'm your new neighbour on the left. I'm the one who's just bought the little house with a bedroom, kitchen, bathroom, living room, toilet and broom cupboard."

At which, there on the doorstep, the old lady clasped her hands, gazed at me with great compassion and began a long lament.

"Alas, my poor m'sieur, yer quite through yer luck today! Oh, 'n it's a pitiful thing to see, a kin' young man like yerself! Well, Lord willin' yer'll come through it sum'ow... Long as yer've life, yer've 'ope, as they say, an' long as yer 'ave yer 'ealf..."

Hearing this, I began to grow nervous:

"But really, my dear madame, can you at least

tell me what's wrong? Everyone I talk to about the house—"

But the old lady interrupted me instantly:

"I 'ope yer'll excuse me, my dear m'sieur, but I've me roast in the oven... I must be off 'n see it don't burn!"

Bam! She too slammed the door in my face.

This time, I was angry. I went back to the solicitor and said to him:

"Now you'd better tell me what's so amusing about my house, so I can join in the fun. And if you don't wish to tell me, rest assured that I will split your head in two!"

With these words, I raised his big glass ashtray, menacingly. This time the bloke stopped laughing:

"Now, now, gently does it! Calm yourself, my dear monsieur! Please put that down and take a seat."

"First you can do some explaining!"

"But of course, I'll explain. After all, now you've signed the contract, I might as well tell you... the house is haunted!"

"Haunted? Haunted by whom?"

"By the witch in the broom cupboard!"

"Couldn't you have told me earlier?"

"Not at all! If I'd told you, you wouldn't have wanted to buy the house, and I wanted to sell it. He he he!"

"Enough giggling, or I'll crack your head open!"

"All right, all right..."

"But tell me, now I think about it: I looked into the broom cupboard, less than fifteen minutes ago... I didn't see any witch in there."

"That's because she's not there in the daytime. She only comes out at night."

"And what does she do there, during the night?"

"Oh, she keeps to herself, she doesn't make any noise, she just stays there, quite well behaved, in her cupboard... only beware! If you should have the misfortune to sing:

Witchy witch, beware,
Watch out for your derrière!

"Then she'll come out... and it'll be your turn to watch out!"

Hearing this, I leapt to my feet, shouting:

"You idiot! You'd no need to go singing that for me. It would never have crossed my mind to sing such tosh. Now, I'll never be able to get it out of my head!"

"That's the idea! He he he!"

And, just as I lunged for his neck, the solicitor escaped through a small door hidden behind him.

What could I do? I went back home, thinking:

"After all, I only have to be a little careful... Let's try to forget that idiotic rhyme!"

Easier said than done, for words like those are not easily forgotten. For the first few months, of course, I was on my guard. Then, after a year and a half, I was comfortable in the house, I had grown used to it, it was familiar... So I began to hum the tune during the day, when the witch wouldn't be there... And then, outside, where I was in no danger... And then I started singing it at night, in the house—but not the whole rhyme! I only sang the beginning:

Witchy witch, beware...

And then I would stop. When I did that, I sometimes thought I saw the door to the broom cupboard begin to shake... But since I always stopped at that point, the witch couldn't do a thing. Realizing this, I began to sing a little bit more each day: *Watch out...* then *Watch out for...* and then *Watch out for your de...* and even *Watch out for your derri...* I would stop just in time! There was no doubt about it, the cupboard door was shaking, rattling, on the verge of coming open... The witch must have been furious in there!

This little game went on until last Christmas. That night, after having Christmas Eve supper with friends, I came home, a little tipsy, just as the clock was striking four in the morning, singing to myself all the way:

> *Witchy witch, beware,*
> *Watch out for your derrière!*

Of course, I wasn't running any real risk, for I was outside the house. I reached the high street: *Witchy witch, beware...* I stopped outside my front door: *Watch out for your derrière!...* I took the key from my pocket: *Witchy witch, beware...* I was still in no danger... I slid the key into the lock: *Watch out for your derrière...* I turned it, went in, took the key out again, closed the door behind me, went down the corridor, towards the stairs...

Witchy witch, beware,
Watch out for your derrière!

Blast! I'd done it now! This time I'd sung the whole rhyme! And then I heard, very close by, a shrill, mean, nasty little voice:

"Oh really! And why exactly should I be looking out for my *derrière?*"

It was her. The broom-cupboard door was open and the witch was standing on the threshold, her right hand on her hip and one of my brooms held in her left. Naturally, I tried to explain:

"Oh, I'm so sorry, madame! A moment of distraction... I forgot... I mean, I meant to say... I hummed it without thinking..."

She chuckled softly:

"Without thinking? Liar! For two years now, that song is all you've been thinking about. You made a fine fool of me, didn't you, stopping every time just before the last word, the last syllable, even! But I said to myself: 'Patience, my pretty! One day, he'll spit it all out, his little sing-song, from start to finish, and when that day comes it will be my turn to have some fun...' And here we are. The day has come!"

I fell to my knees and began to plead:

"Have pity, madame! Don't hurt me! I didn't mean to offend you: I actually really like witches. Some of my best friends are witches! My poor mother was a witch! If she weren't dead, she could tell you herself... And besides, today is Christmas Day! Little Baby Jesus was born tonight. You can't make me disappear today, of all days..."

The witch replied:

"Taratata! I won't listen to a word! But since you've got such a ready tongue, I'm going to set you a challenge: you have three days in which to ask me for three things. Three impossible things! If I can give you all three of them, you're mine. But if I am unable to give you any one of the three things, I shall disappear for ever and you'll never see me again. So, I'm listening!"

Playing for time, I replied:

"Hm, I don't know... I've no idea... I'll have to think about this one... Can I have today to think about it?"

"Fine," said the witch. "I'm in no rush. See you in the evening!"

And she disappeared.

Sitting in thought for several hours, I cudgelled,

wrestled with and generally racked my brains—when suddenly I remembered that my friend Bashir had two little fish in a bowl, and that he had said these two little fish were in fact *magic* fish. Without losing another second, I raced down rue Broca to go and ask Bashir:

"Have you still got your two little fish?"

"Yes. Why?"

"Because there's a witch in my house, a really old, wicked witch. I have to ask her for something impossible by tonight. If I don't, she'll spirit me away. Do you think your little fish might give me an idea?

"Sure they will," said Bashir. "I'll go and get them."

He went into the back of his father's shop and came back with a bowl full of water in which two little fish were swimming, one red and the other yellow with black spots. They really did look like magic fish.

"Now, speak to them!"

"I can't!" Bashir replied. "*I* can't talk to them; they don't understand French. We need an interpreter! But don't worry—we have one here."

And my friend Bashir began to sing:

Little mouse
Little friend
Will you come this way?
Speak to my little fish
And you shall have a tasty dish!

Hardly had he finished singing when an adorable little grey mouse came trotting out onto the counter, sat down on her little bottom beside the fishbowl and gave three tiny squeaks, like this:

"Eep! Eep! Eep!"

Bashir translated:

"She says she's ready. Tell her what happened to you."

I bent down and told the mouse everything: all about the solicitor, the house, the neighbours, the cupboard, the song, the witch and the challenge she had set me. After listening to me in silence, the mouse turned to the little fish and said to them in her language:

"Eepi eepeepi peepi reepeeteepi..."

And on like that for another five minutes.

*

Once they too had listened in silence, the fish exchanged glances, consulted, whispered in each other's ears, and then finally, the red fish rose to the surface of the water and opened his mouth several times, making a tiny, almost inaudible sound:

"Po—po—po—po..."

And so on, for nearly a minute.

When that was done, the little mouse turned back to Bashir and began squeaking again:

"Peepiri peepi reepipi."

I asked Bashir:

"What is she saying?"

He replied:

"This evening, when you see the witch, ask her for jewels made of rubber that shine like real gemstones. She won't be able to bring you any."

I thanked Bashir. Bashir dropped a few water-fleas into the bowl for the fish to eat and gave the mouse a round of salami, and I left the shop to go back home.

The witch was waiting for me in the corridor:

"So? What will you ask me for?"

I replied, confidently:

"I want you to give me jewels made of rubber that shine like real gemstones!"

But the witch began to laugh:

"Haha! You didn't think up that one by yourself! But never mind, here they are."

She rummaged about inside her bodice and pulled out a fistful of jewellery: two bracelets, three rings and a necklace, all shining just like gold, glittering like diamonds, in all the colours of the rainbow—and soft as the rubber in your pencil case!"

"See you tomorrow for your second request," said the witch. "And this time, try to make it a little more challenging!"

And—*pouf!* She disappeared.

The following morning, I took the jewellery with me to a friend who is a chemist, and asked him:

"Can you tell me what this substance is?"

"Give me a minute," he said.

And he locked himself up in his lab. After an hour, he came back out, saying:

"This is quite extraordinary! They're made of rubber! I've never seen such a thing. May I keep them?"

I left the jewellery with him and went back to see Bashir.

"The jewellery was no good," I told him. "The witch brought them to me straight away."

"In that case, we'll have to try again," said Bashir.

He went back to get the fishbowl, set it on the counter and began to sing once more:

Little mouse
Little friend
Will you come this way?
Speak to my little fish
And you shall have a tasty dish!

The little mouse ran out, I told her what had happened, she translated, then listened to the reply from the fish and transmitted it to Bashir:

"Peepi pirreepipi ippee ippee ip!"

"What does she say?"

And Bashir translated for me, again:

"Ask the witch for a branch from the macaroni tree, and replant it in your garden to see if it will grow!"

That very evening, I said to the witch:

"Bring me a branch from the macaroni tree!"

"Haha! That's not your own idea either! But no matter: here you go."

And *pouf!* She pulled a magnificent branch of flowering macaroni out of her bodice, with twigs made of spaghetti, long noodle leaves and pasta-shell flowers. It even had little seeds shaped like alphabet pasta!

I was quite amazed, but even so, I couldn't let the witch off so easily:

"That isn't a branch from a real tree—it won't grow!"

"That's what you think," said the witch. "Just plant it out in your garden and you'll see. Catch you tomorrow evening!"

Without further ado, I went into the garden, dug a small hole in a flower bed, planted the macaroni branch in it, watered it and went to bed. The following morning, I went downstairs to look. The branch had grown huge: it was almost a whole new tree, with several fresh branches and twice as many flowers. I gripped it with both hands and tried to pull it up... but I couldn't! I scratched at the ground around the trunk and I saw that it was being held tight to the ground by hundreds of its own tiny vermicelli roots...

This time I was desperate. I didn't even feel like going back to Bashir. I wandered around like a soul in pain, and I'm sure I saw people whispering when they saw me go by. I knew what they were saying, too!

"That poor young man—just look at him! It's his last day on this earth, you can see straight away. The witch will surely carry him away tonight!"

On the stroke of midday, Bashir gave me a call:

"So? Did it work?"

"No, it didn't. I am lost. The witch is going to carry me off tonight. Goodbye, Bashir!"

"Not at all, nothing is lost. Why are you going on like this? Come round here right now and we'll ask the little fish!"

"What for? It's no good."

"And what good is doing nothing? I'm telling you, come to my place right away! It's shameful to give up like this!"

"All right, if you wish, I'm coming..."

And I went back to Bashir's house. When I got there, everything was ready: Bashir, the bowl with the little fish and the little mouse sitting beside it.

For the third time I told my story, the little mouse translated it, the fish discussed it at length, and this time it was the yellow fish that came to the surface to speak in a series of gulps:

"Po—po—po—po—po—po—po..."

He went on for nearly a quarter of an hour.

Next the mouse turned back to us and made a whole speech, which took a good ten minutes.

I asked Bashir:

"What on earth can they be going on about, this time?"

Bashir told me:

"Listen, and do pay attention because this is not so simple. This evening, when you get back home, ask the witch to bring you the *hairy frog*. She will be very embarrassed, for the hairy frog is in fact the witch herself! The witch is no more, no less, than the hairy frog in human form. Now, one of two things should happen: either she won't be able to bring you the hairy frog, in which case she will have to leave your house for ever—or she will decide to show you the frog anyway, and to do that, she will have to transform herself back into it. As soon as she has turned into the hairy frog, you must catch her and tie her up good and tight with thick string. Then she won't be able to grow back into the witch again. After that, you must shave her hair off and then you'll be left with a perfectly inoffensive, ordinary frog."

Now I began to feel hopeful again. I asked Bashir:

"Could you sell me the string?"

Bashir sold me a ball of tough string. I thanked him and went back home.

Come evening, the witch was there, waiting for me:

"So, my pretty, has the time come for me to spirit you away? What are you going to ask me for now?"

I made sure that the string was nice and loose in my pocket, and then I replied:

"Bring me the hairy frog!"

Now the witch stopped laughing. She gave a shriek of rage:

"What? What did you say? You didn't think this one up by yourself either! Ask me for something else!"

But I stood firm:

"Why should I ask for something else? I don't want anything else; I want the hairy frog!"

"You have no right to ask me for that!"

"Is it that you *can't* bring me the hairy frog?"

"Of course I can, but this isn't fair!"

"So you don't want to bring it to me?"

"No, I don't want to!"

"In that case, go back where you came from. This is my home!"

At that, the witch began to screech:

"Oh, it's like that, is it? In that case, here you go, since this is what you want; you shall have your hairy frog!"

And before my eyes she began to grow smaller, to dwindle, shrivel and shrink as if she were melting away, so much and so completely that within five minutes there before me was nothing but a fat, green frog with a thick crop of hair on her head, hopping around on the floor and croaking as if she had hiccups:

"Ribbit, ribbit! Ribbit, ribbit!"

I jumped on her right away and pinned her to the ground. Pulling the string from my pocket, I took her and trussed her up like a chunk of salami... She wriggled, almost strangling herself; she did her best to grow back into the witch... but the string was too tight! Her eyes bulged furiously at me, while she croaked desperately:

"Ribbit, ribbit! Ribbit! Ribbit!"

Without hesitating, I carried the frog into the bathroom, soaped her up and shaved her hair off, after which I untied her, ran a little water into the bath and left her to spend the night there.

The next morning, I took her to Bashir in a small bowl with a tiny ladder inside it, so she could help forecast the weather for him (she would climb up the ladder if good weather was coming and sit at the bottom if the forecast was bad). Bashir thanked me and put the new bowl on a shelf, next to the one with the little fish.

Since that day, the two fish and the frog have not stopped talking to each other. The frog says: "Ribbit, ribbit!" and the fish: "Po—po!" and they go on like that for days on end!

One day I said to Bashir:

"How about you call the mouse in, so we can find out what they're saying to each other?"

"Sure, if you like!" Bashir said.

And once more, he sang:

> *Little mouse*
> *Little friend*
> *Will you come this way…*

When the mouse came, Bashir asked her to listen and translate. But this time, the mouse refused point blank.

"Why won't she translate?" I asked.

Bashir replied:

"Because it's nothing but swearing!"

So now you know the story of the witch in the broom cupboard. And now, if you come and visit me in my little house, whether by day or by night, you can quite safely sing:

Witchy witch, beware,
Watch out for your derrière!

I promise nothing will happen to you!

fin

The Love Story of a Potato

There was once a potato—a common potato, such as we see every day—but this one was eaten up with ambition. Her lifelong dream was to become a French fry. And this is probably what would have happened to her, had the youngest boy in the house not stolen her from the kitchen.

As soon as he had his booty safely in his bedroom, the little boy pulled a knife from his pocket and set about carving the potato. He began by giving her two eyes—and at once the potato could see. After which he gave her two ears—and the potato was able to hear. Finally, he gave her a mouth—and the potato was able to speak. Then the boy made her look at herself in a mirror, saying:

"See how beautiful you are!"

"How dreadful!" replied the potato. "I am not

the least bit beautiful. I look like a boy! I was much happier before."

"Fine, okay then!" replied the little boy, annoyed. "If that's how you see it..."

And he threw the potato in the bin.

Early the next morning, the bin was emptied and, later that day, the potato was dumped along with a great heap of other rubbish, in the middle of the countryside.

"An attractive region," she said, "and very popular at that! What a collection of fascinating people there are here... Now, who can that be, looking rather like a frying pan?"

It was an old guitar, nearly split in half, with only two strings left intact.

"Hello there, madame," said the potato. "It seems to me, from your appearance, that you must be a very distinguished person, for you bear a marvellous resemblance to a frying pan!"

"You are very kind," said the guitar. "I do not know what a frying pan would be, but I thank you all the same. It's true that I'm not just anybody. My name is Guitar. And yours?"

"Well, my name is Maris Piper. But you can call me

Potato for, from today, I shall count you an intimate friend. Because of my beauty, I was selected to become a French fry, and I should have become one had I not suffered the misfortune of being stolen from the kitchen by the youngest boy in the house. What is worse, having stolen me, the scoundrel completely disfigured me with these pairs of eyes and ears and this awful mouth..."

And the potato began to cry.

"Now, now, don't cry," said the guitar. "You are still very elegant. And besides, this means you can speak..."

"That's true," agreed the potato. "It's a great consolation. In the end—to finish my story—when I saw what that little monster had done to me, I was furious, and I wrenched the knife right out of his hands, cut off his nose and ran away."

"Well done, you!" the guitar responded.

"Don't you think?" said the potato. "But, what about you? How do you come to be here?"

"Well," replied the guitar, "for many years I was best friends with a handsome young boy, who loved me dearly. He used to bend over me, take me in his arms, caress me, strum me, pluck the strings on my belly while singing such delightful songs to me..."

The guitar sighed, then her voice grew bitter and she went on:

"One day he came back with a strange instrument. This one was also a guitar, but made of metal, and oh so heavy, vulgar and stupid! She took my friend from me, she bewitched him. I am sure he didn't really love her. He never sang her any tender songs when he picked her up—not one! He used to pluck furiously at her strings and give savage howls and roll about on the ground with her—you would have thought they were fighting! Besides, he didn't trust her! The plain proof is that he kept her tied up on a leash!"

In fact, what had happened was that the handsome young man had bought an electric guitar, and what our guitar had taken for a leash was in fact the wire that connected the new guitar to the electricity.

"Anyway, however it happened, she stole him from me. After only a few days he only had eyes for her, he no longer looked at me at all. And when I saw that, well, I preferred to leave him..."

The guitar was lying. She had not left of her own accord; her master had thrown her out. But she would never have admitted that.

In any case, the potato hadn't understood a word.

"What a beautiful story!" she said. "How moving! I'm quite beside myself. I knew we were made to understand each other. Besides, the more I look at you, the more I feel you look like a frying pan!"

But while they were chatting like this, a tramp going by on the high road heard them, stopped and listened harder.

"Now this is no ordinary how-d'ye-do," he thought. "An old guitar telling her life story to an old potato, and the potato answering. If I can do this right, I'll be a rich man!"

He found a way into the wasteland, picked up the potato and put her in his pocket, then he grabbed the guitar and took the two friends with him to the next town.

This town had a large central square, and in the square there was a circus. The tramp went and knocked on the circus ringmaster's door.

"Mista Ringmaster! Mista Ringmaster sir!"

"Hmph? What? Come in! What do you want?"

The tramp stepped into the caravan.

"Mista Ringmaster, I have a talking guitar!"

"Hmph? What? Talking guitar?"

"Yes yes, Mista Ringmaster! And a potato that answers it back!"

"Hmph? What? What is this story? Are you drunk, my friend?"

"No, no, I'm not drunk. Please just listen!"

The tramp put the guitar on the table, then took the potato from his pocket and put them next to each other.

"Now, hop to it. Talk, you two!"

Silence.

"Talk, I tell you!"

More silence. The Ringmaster's face flushed an angry red.

"Tell me, my friend, did you come here purely to make a fool of me?"

"Of course not, Mista Ringmaster! I'm telling you, they do talk, both of them, to each other. Just now, they're being difficult so as to annoy me, but..."

"Get out!"

"But when they are alone..."

"I said: get out!"

"But Mista Ringmaster..."

"Hm? What? You haven't left yet? Very good, I shall throw you out myself!"

The ringmaster caught the tramp by the seat of his pants and—therr-whumpp!—he tossed him out. But at that very moment, he heard a great burst of laughter behind him. Unable to hold her tongue any longer, the potato had just said to the guitar:

"Hey, do you think we fooled him? He he he!"

"And how! We fooled him good and proper!" the guitar was saying. "Ha ha ha!"

The ringmaster whirled around:

"Well I never, how about that! The old drunk was telling the truth. You can talk, both of you!"

Silence.

"Come on," the ringmaster went on. "There's no point keeping quiet now. You can't fool me any longer: I heard you!"

Silence.

"That is a pity!" the ringmaster said then, with a cunning expression. "I had a rather exciting proposal for you. An artistic proposal!"

"Artistic?" asked the guitar.

"Shut up!" hissed the potato.

"But I adore art!"

"Now we're getting somewhere!" said the ringmaster. "I can see that you're a sensible pair.

And indeed, you will have work, both of you—oh yes you will. You will become stars."

"I'd rather become a French fry," objected the potato.

"A French fry? You—with your talent? That would be a crime! Would you really prefer to be eaten than to be famous?"

"What do you mean, 'eaten'? Do people eat French fries?" asked the potato.

"Do we *eat* French fries? Of course we eat them! Why do you think we're always frying more?"

"Really? I didn't know!" said the potato. "Well, if that's how things stand, then fine. I'd rather become a star."

A week later, all over the town, big yellow posters appeared on which were written:

THE FABULOUS CIRCUS OF WHATSIT

See clowns! See acrobats!
Bareback riders! Trapeze-artists!
See tigers, ponies, elephants, fleas!
And, in their world premiere show:
NOÉMIE, the performing potato
And AGATHE, the guitar who plays herself!

The big top was full on the new show's first night, for nobody in that part of the world had seen anything like it before.

When their turn came, the band played a military march while the potato and the guitar stepped bravely into the ring. To start with, the potato introduced their number. Then the guitar played a difficult piece by herself. Then the potato sang a song, accompanied by the guitar, who sang a harmony while playing herself at the same time. And then, the potato pretended to sing a wrong note and the guitar pretended to catch her out. The potato pretended to get angry and they both pretended to have a big argument, to the great delight of the audience. Finally, they pretended to make up and be friends again and they sang their last song together.

The potato and the guitar were a huge success. Their act was recorded for radio and for television and, soon, people were talking about it all over the world. Having seen it on the news, the Sultan of Bakofbiyondistan flew over that afternoon in his private jet, to see the ringmaster.

"Hello, Mister Ringmaster."

"Hello, Mister Sultan. What can I do for you?"

"I should like to marry the potato."

"The potato? Now, look here, she's not a person!"

"Very well, I'll buy her."

"But she's not an object either... She speaks, she can sing..."

"Very well, I'll take her from you!"

"But you've no right to do that!"

"It's my right to do anything I please, for I have oodles of money!"

The ringmaster realized he should try to be a little cleverer.

"You will cause me great sadness," he said, sobbing. "I love that potato, I've grown attached to her..."

"And how I sympathize!" said the Sultan, with just a hint of sarcasm. "In that case, I can offer you a caravan full of diamonds for her!"

"Just the one caravan?" asked the ringmaster.

"Two, if you prefer!"

The ringmaster wiped away a tear, blew his nose loudly, then added in a wobbly voice:

"I feel, if you were to go as far as three caravans..."

"Done! Three it shall be, and let that be an end of it."

The next day, the Sultan flew back to his sultanate, taking the potato with him, and also the guitar, for the two old friends were determined to stay together. That week, a popular weekly magazine published a photograph of the brand-new couple with the following front-page headline:

WE LOVE EACH OTHER

In the weeks that followed, the same magazine published more photos, and the headlines changed accordingly. In order of appearance, they went like this:

WILL THE GOVERNMENT DARE TO STOP THEM?

WILL IT BREAK THE POTATO'S HEART?

POTATO SAYS, WEEPING: THIS CAN'T GO ON!

GUITAR SAYS: I'D RATHER GO!

AND STILL THEY ARE IN LOVE!

LOVE CONQUERS ALL

And beneath that last headline followed more photographs—from the wedding of the Sultan and the potato. Only a week later, the newspapers were full of other news, and soon, everyone had forgotten all about the love story of the Sultan, the potato and the guitar.

The Cunning Little Pig

Once upon a time a mummy god was sitting in a big armchair, darning socks, while, sitting at the dinner table, her young god was finishing his homework.

The young god worked away in silence. And when he was finished, he asked:

"Mummy, can I be allowed to make a world?"

Mummy God looked over at him:

"Have you finished all your homework?"

"Yes, Mummy."

"Have you learnt your lessons?"

"Yes, Mummy."

"Good boy. Then, yes, you may."

"Thanks Mummy."

The young god took a piece of paper and some coloured pencils and set about making his world.

*

First, he created the sky and the earth. But the sky was empty and so was the earth, and both were covered in darkness.

So the young god created two lights: the Sun and the Moon. And he said aloud:

"Let the Sun be the man and the Moon be the lady."

So the Sun became the man and the Moon the lady, and they had a little daughter, who was called Dawn.

Next the young god made plants to grow on the earth and seaweed to grow in the sea. Then he made animals to live on the earth: some to crawl on the ground, some to swim in the sea and some to fly in the air.

Next he created people, the most intelligent of the animals to live on his earth.

When he had made all this, the earth was full of life. But in comparison, the sky looked rather empty. So the young god shouted as loudly as he could:

"Which of you animals wants to come and live in the sky?"

Everybody heard, except for the little pig, who was busy eating acorns. For the little pig is so greedy that he doesn't notice anything when he's eating.

Now all the animals that wanted to live in the sky responded to the young god's call: the ox replied, and the bull and the lion; the scorpion and the crab, whose name was Cancer; the swan and all the fish; both centaurs responded, one of them being the archer Sagittarius; both bears were there, the Little and the Great; so were the whale and the hare; the eagle and the dove; the dragon, the snake, the lynx and the giraffe all responded; there was a little girl who was called Virgo; there was a whole bunch of Greek letters, and even a few objects responded, such as Libra, the weighing scales.

This crowd of creatures came together and began to shout:

"Me! Me! Me! I want to live in the sky!"

So, the young god picked them all up, one by one, and stuck them up in the heavenly vault, with the help of those big silver drawing pins that we call stars. It did hurt them a little, but they were so happy to be living in the sky that they didn't give the star pins a second thought!

When the whole exercise was over, the sky was studded with creatures, while the stars shone in all their magnificence.

"This is all very pretty," said the Sun, "but when I rise in the morning, I'll grill them alive!"

"That's true," admitted the young god, "I hadn't thought of that!"

He pondered for a moment, then he said:

"Right, in that case, it's quite simple: every morning, young Dawn will get up before her father the Sun and take down everyone who lives in the sky. And every evening, when the Sun has set, she will pin everyone back up there!"

And this is what they did. This is why, every morning, the stars disappear, only to return again at the end of the day, after dark.

Everything being now thoroughly organized, the young god looked down on his world with satisfaction.

"You know," said Mummy God, "it's just about time for bed. You have school tomorrow!"

"I'm coming, Mummy," said the young god.

And he was about to get up when he heard a loud noise. It was the little pig racing in, as fast as he could, all out of breath and shouting as loudly as he could:

"What about me, then? What about me?"

"Well, what about you?" the young god asked.

"Why can't I go and live in the sky too?"

102

"Why didn't you ask me before?"

"No one told me you had to ask!"

"What do you mean, no one told you!" exclaimed the young god. "Didn't you hear, when I called for volunteers?"

"No, I didn't hear anything."

"What were you up to, that you didn't hear?"

"I think," said the little pig, blushing, "that I was eating acorns..."

"Well, hard luck for you!" said the young god. "If you weren't such a greedy guts, you might have heard me. I did shout very loudly!"

At this, the little pig began to sob:

"Oh pleeease, Mister Young God, sir! You can't leave me behind like this. Can't you squeeze me in somewhere? Tell the others to shuffle up a bit... If need be, you could pin me up on top of them! But do something, please, I don't mind kissing your feet..."

"I can't!" said the young god. "First because there's no more space, you can see that for yourself. The others can't squeeze together any closer. Besides, there aren't any more stars to pin you up there. And lastly, I haven't time: my mother has been calling me for a good minute already!"

With these words, the young god stood up from the table and went off to bed. Within ten minutes, he was asleep, and had quite forgotten about the brand-new world he had created. Meanwhile, the little pig was rolling about on the ground, sobbing:

"I want to be up in the sky! I want to live in the sky!"

But when he grew tired of rolling on the ground, he stopped and looked around, and realized the others had left him all by himself. So he settled down on the ground, laid his snout on his front trotters and began to grizzle:

"I knew they didn't like me! Nobody likes me. They all hate me—even that god! He's taken against me. He called while I was eating on purpose, so that I wouldn't hear. And he made sure to fill up the sky with everyone else double quick, so that I'd be too late. And what's that supposed to mean: that there aren't any stars left for me? Couldn't he make any more, huh? Oh, but I shall have my revenge! This isn't the end of the story! So he says there aren't any stars left for me; well, we shall see about that!"

He got up and trotted away in search of young Dawn.

Dawn had just got up, for the night was nearly over, and she was brushing her hair, getting ready to go, when the little pig trotted into her room:

"My poor little Dawn!" he said, with a sorrowful expression. "How unhappy you must be!"

"Unhappy, me? Not at all!"

"Oh, but you must be unhappy!" said the little pig. "Your parents are so hard on you!"

"Hard, my parents? Why do you say that?"

"Why? Isn't it hard to force a child of your age to get up before daylight in order to pull down all the stars in the sky? And to make her stay up until dark so as to pin them all up again? I'm shocked every time I think about it!"

"Listen," little Dawn said, "you mustn't let yourself be so easily shocked! My work is rather good fun, you know... It doesn't bother me. And besides, it isn't my parents' fault! It's the young god who ordered this!"

"Let's not even mention the young god," said the little pig, bitterly.

"Oh, I'm sorry. Have I upset you?"

"Forget it, it's nothing... You know, I only want one thing in life, and that's to serve you. But if you hate me too much to accept my offer, well then..."

"But I don't hate you!" little Dawn protested. "What is it that you want, exactly?"

"Oh, I don't want anything for myself. I simply thought to suggest..."

"Spit it out, then; what is it you'd suggest?"

The little pig lowered his voice:

"Well, if you like, I could come with you this morning and help you with your work..."

"Well," said young Dawn, "if that's all it takes to make you happy..."

"But it's not to make *me* happy!" the little pig explained, loftily. "I want to help you—that's all I want to do!"

"All right then. Let's go!"

Dawn put down her hairbrush, picked up a vast sack and slung it over her shoulder, and off they went.

As soon as they had reached the sky, they set to work. The little pig held the sack open while Dawn tossed the stars down into it pell-mell, on top of each other. As they were unpinned, the animals living up in the sky began to come down to earth where they would spend the day.

"This is wonderful!" said young Dawn. "I'm going twice as fast as usual! Thank you so much, little pig!"

"It's nothing, nothing at all!" puffed the little pig, chuckling to himself.

Now, just as Dawn was tossing the Little Bear's stars into the open sack, the little pig jumped at the most beautiful one—the Pole Star, the one that shows which way is north. He caught the star in mid-air, swallowed it up like a truffle and ran away as fast as his trotters could carry him.

"Little pig! What on earth are you doing?" called young Dawn, after him.

But the little pig pretended not to hear her. He sped back to earth at top pig-speed and very soon vanished from view.

What could she do? Dawn would have gone after him there and then, but first she absolutely had to finish taking the stars down from the sky, for the horizon was already growing paler in the east. She got back to her work and only when she had finished did she set out in search of the Pole Star.

From sunrise until midday, she criss-crossed Asia. But nobody there had seen the little pig. From midday until four o'clock, she combed the continent of Africa.

But the little pig had not been seen there either. From four o'clock, she searched all over Europe.

Meanwhile, knowing Dawn would be looking for him, the little pig had taken refuge in France, in a city called... —Well, what was that city called?—Oh yes! A city called Paris. And while scurrying all over Paris, he happened to turn into a street called... —What *was* that street called, now?—Yes, of course: rue Broca! And, on reaching a shop at number 69 rue Broca, the pig vanished into its open door. This was the cafe-grocer's belonging to...—Oh dear, my memory! Who did it belong to?—Oh, yes. To Papa Sayeed!

Papa Sayeed was not there. Nor was Mama Sayeed. Both of them were out, I don't know why. What's more, their eldest daughter Nadia had been stolen away by the wicked witch of rue Mouffetard, and her younger brother Bashir had gone to save her. So now the only people left to look after the shop were the Sayeeds' two youngest daughters: Malika and Rashida.

There the two girls were, enjoying the early-afternoon peace and quiet, when a gust of wind suddenly blew through the shop and, along with it tumbled a little pig—a rather pretty little pig, in fact, whose tightly stretched skin gave out a delicate pink glow (from the star that was glowing inside his tummy). The little pig begged them, breathlessly:

"Save me! Please, save me!"

"What should we save you from?" asked Malika.

"From a little girl! From young Dawn! She's coming after me! She wants to kill me! And eat me whole!"

"No way!" gasped Rashida.

"She does, she does! She's been chasing me since morning! If you don't hide me, she will eat me up!"

And fat tears began to roll down the little pig's cheeks.

The two girls looked at each other.

"Poor thing," said Malika.

"We must do something!" Rashida decided.

"What if we hide him in the cellar?" suggested Malika.

"That's a good idea!"

They sent the little pig down into the cellar and were about to close the trapdoor when he stopped them for a moment:

"Now, if anyone asks for me, you haven't seen me. Understood?"

"All right!" said Malika.

"Oh, and I was forgetting: young Dawn will doubtless tell you some yawn of a shaggy-dog story about some star she'll say I've eaten... Obviously, it's total nonsense: little pigs do not eat stars. I hope you won't believe her for a moment..."

"Of course not!" said Rashida.

"And one more thing! Don't tell your parents about me, it's better if you don't... Parents, you know, they're rather stupid, they don't understand how life works..."

"Okay!" said the two girls, together.

And they let the trapdoor fall closed. Then they looked at each other:

"Why doesn't he want us to tell our parents?" whispered Malika, anxiously. "There's something funny about him!"

"And why does he glow in the dark like that?" asked Rashida. "Did you see him there, in the cellar, while he was talking to us? He looked like a lamp with a pink lampshade!"

Malika scrunched her nose up: she was thinking.

"Perhaps his story about the star is true, after all..."

"But then, are we wrong to hide him?" asked Rashida, very worried.

"Never mind." said Malika. "We should have thought of that before! Now we've taken him in, we can't betray him."

At about five o'clock that afternoon, young Dawn walked into the shop.

"Hello, young ladies! You wouldn't, by any chance, have seen a little pink pig today, would you?"

"Pink all over and glowing like a nightlight?" asked Malika.

"Just like that!"

"No, we haven't seen him!"

"In that case, I'm sorry to disturb you," said young Dawn. "Goodbye, ladies!"

And she left the shop. But five minutes later, she was back:

"Forgive me, ladies. It's about this little pig... If you haven't seen him, how do you know that he glows?"

"It's because he has eaten a star," replied Rashida.

"Indeed he has! Have you seen him, then?"

"No, never!"

"Oh. Right."

And young Dawn left the shop for the second time. Hardly had she stepped outside when she stopped and frowned, then went back into the shop:

"Forgive me, ladies, it's me again... Are you really completely sure that you haven't seen the little pig?"

"Oh yes, quite sure! Absolutely sure!" chorused Malika and Rashida, blushing as pink as pink roses.

Young Dawn gazed at them doubtfully, but since she had no proof, she did not dare challenge them again and so off she went once more, for good this time.

At six o'clock in the evening, Papa and Mama Sayeed came home. They asked the girls:

"Any news from the shop today?"

"Yes!" they said. "Nadia was taken away by the wicked witch."

"Oh? And then?"

"Then Bashir went to save her."

"Oh, good! Anything else?"

"No, nothing else..."

"Very good. Go and have your tea."

A few hours later, the day was almost over. Poor Dawn had searched the whole world but had no luck, and already it was time for her to start pinning the animals who lived in the sky back up there. She picked up her sack of stars, called all the heavenly animals and began to pin them all up again. When she got to the Little Bear, she pinned him up as best she could with the stars she had left, and she was about to go on, when Little Bear stopped her:

"Well? What about my Pole Star? You're forgetting my Pole Star!"

"Drat!" hissed young Dawn into Little Bear's ear. "I think I've lost it. But don't tell anyone. I promise I'll find it for you before tomorrow evening..."

But the Little Bear didn't hear very well in that ear. She began to cry:

"Waaah! My Pole Star! Waah! I want my Pole Star! Waaaaah! The little girl has lost my Pole Star..."

She was making such a racket that the Moon hurried over:

"What's all this? What's going on?"

Very ashamed, young Dawn told her mother what had happened.

"Why didn't you tell me before?"

"I didn't dare, Mama... I thought I could find the star by myself."

"Oh well, that wasn't very clever, was it! Now we shall have to tell your father! And he does not like being woken up, does your father, once he has gone to bed!"

Poor young Dawn finished her work with her

mother helping, sniffing as she went. When they were finished, they went to wake up the Sun.

That night—a beautiful, clear night—there was no Pole Star, but instead a great black space in the sky. And a great many ships that had set out for America ended up in Africa or even in Australia, because they had lost track of where north was.

"Oh, very clever!" grumbled the Sun in a thunderous voice, throwing flames in all directions. "What in heaven can I have done to deserve such a little idiot... I don't know what's stopping me from—"

"Now, now, don't get so worked up," said the Moon, impatiently. "What good will it do?"

"True," admitted the Sun. "But all the same."

Then, turning to young Dawn, he asked:

"Look, what is it that happened, exactly? Tell me everything."

And, when young Dawn had finished her tale, he said:

"That little pig is doubtless hiding at Papa Sayeed's shop. Those little girls must have hidden him. Quick, bring me my great black cloak, my black hat, my black scarf, my black mask and my dark glasses, and I'll be there in a flash."

The Sun put on his great black cloak, his black hat, his black scarf, his black mask and his dark glasses. Dressed like this, no one could tell that he was in fact the Sun. He went down to earth and straight away to see Papa Sayeed.

When he stepped into the shop, Papa Sayeed asked: "What will it be for monsieur?"

"Nothing," said the Sun. "I would like to talk to you."

Hearing this, Papa Sayeed took him for a door-to-door salesman:

"In that case," he said, "you can come back tomorrow! Why do you always come at this time? You can see that I have customers to serve!"

"I am not who you think I am," said the Sun. "I have come to look for the little pig that has eaten the Pole Star."

"What kind of a tall tale is this? There's no little pig here!"

"And I," said the Sun, "I am certain that there is. Your children let him in."

Papa Sayeed called in his four children, who were watching television:

"Now, what's this story I'm hearing about you? Have any of you four seen a little pig today?"

Nadia said: "I wasn't here during the day—the witch stole me away."

"Me neither," said Bashir, "I went to save Nadia."

But Malika and Rashida stood there in silence, looking at the floor. Papa Sayeed asked:

"And what about you two, now? Have you seen a little pig?"

"A little pig?" asked Malika, in a small voice.

"A little pig?" echoed Rashida.

Papa Sayeed lost his temper.

"Yes, a little pig! Not a hippopotamus, to be sure! Have you both gone deaf?"

"Have you seen a little pig?" Malika asked Rashida.

"Me? Oh no!" Rashida replied. "And you? Have you seen one, a little pig?"

"No, me neither. Not one little pig..."

"Really!" said the Sun. "Are you sure? A little pig, green all over, being chased by an old gentleman with a wooden leg?"

"That's not right!" said Malika indignantly. "He was pink!"

"Besides," added Rashida excitedly, "it wasn't an

old gentleman following him: it was a little girl! And she didn't have a wooden leg!"

Just then, they both went quiet, looked at each other and blushed right up to their ears, realizing that they had given each other away.

"There's our proof!" cried the Sun.

"What does this mean?" Papa Sayeed shouted. "Hiding a little pig in my shop—and what's more, without asking! And trying to lie to me, on top of everything!"

The two little girls began to cry:

"But it's not our fault!"

"We thought we were doing the right thing!"

"He begged us so hard!"

"He pleaded with us!"

"He told us the little girl was going to kill him!"

"Kill him and eat him!"

"Enough lies!" thundered Papa Sayeed. "Come here and let me give you each a good smacking."

But this time, the Sun stepped in.

"Don't smack them, Monsieur Sayeed, I am sure they are telling the truth. I know this little pig: he's a terrible liar and quite capable of telling them all this nonsense."

Then, turning to the two girls, he asked them gently:

"And where have you put him?"

"In the cellar," whispered Malika.

"Would you mind showing me your cellar?" the Sun asked Papa Sayeed.

"Well... I would rather not!" said Papa Sayeed. "I don't much like this kind of thing, myself. And besides, it could cause me problems in the future. I don't even know who you are."

"I am the Sun," said the Sun.

"Then, prove it. Take off your dark glasses!"

"I really can't," said the Sun. "If I take them off, the whole house will catch fire!"

"All right then, keep them on," said Papa Sayeed. "And stay back behind the counter."

He lifted the trapdoor. All the customers in the cafe who had been listening to the conversation crowded over to see. As soon as the trapdoor was raised, a soft pink light shone out.

"He's in there!" cried the Sun.

And, without even asking for the ladder, he stretched out one long, *long* arm, lifted the little pig out by his ear and sat him on the marble shop counter. The little pig wriggled and struggled and yelled as loud as he could:

"Let me go! Let me go! I want to stay here!"

"You can stay where you like," said the Sun, "but I want my star back."

"Star? What star? I don't know any star. I've never even seen a star!"

"Liar!" said the Sun. "I can see it shining right through your tummy!"

The little pig looked down at his tummy, saw the glow and gave up pretending:

"All right—take your star, then." he said. "I don't want anything to do with your star! I never wanted it in the first place! I didn't mean to eat it."

"Don't talk so much," said the Sun, "and spit it out, if you can."

The little pig tried and tried to spit out the star, but he couldn't.

"We'll have to make him throw up," said the Sun.

"I have an idea," said Papa Sayeed.

He took a very big glass and in it he put: coffee, mustard, salt, grenadine syrup, rum, pastis, brandy and beer. The little pig gulped down this horrible mixture, went quite pale and began to vomit up everything inside him—except for the star.

At three in the morning, they sent for a vet, who gave the little pig a purgative meant for horses, hoping they might get the star out by the pig's other end. Between four and five o'clock, the little pig did quite a few things, but still no star came out.

When the clock struck half-past five in the morning, the Sun cried:

"It's too bad! I can't wait any longer. The day is dawning and soon I will have to rise—we shall have to use extreme measures! Monsieur Sayeed, can you bring me a knife?"

Papa Sayeed, who was also getting rather fed up, took out the long knife he used for cutting bunches of bananas. The Sun seized it and, without a moment's hesitation, he sank the knife into the little pig's back, making a large cut. Then he slipped two fingers into the slit, drew out the Pole Star and put it in his pocket. The little pig was weeping, but he didn't make a squeak: he may have been a dreadful liar, but he was, all the same, a very brave little pig.

"Thank you, Monsieur Sayeed," said the Sun. "And please accept my apologies for this sleepless night. Now I have to go, for young Dawn has already begun taking the stars down from the sky. I really don't know how to reward your kindness..."

"Well, I know," said Papa Sayeed. "Just keep shining as hotly as you can, so that my customers are thirsty and my business goes well..."

"Right, it's a deal, I'll do my best!"

Then, turning to the little pig, the Sun added:

"As for your punishment, since you so enjoy eating shiny things, you shall be turned into a piggy bank. You shall keep that slot in your back, Monsieur Sayeed will drop his tips in there, and you won't walk free until you're filled up with coins!"

"Great!" said the little pig. "I'll soon be full!"

"There's an optimist!" said the Sun.

Now, the Sun uttered a magic spell. The little pig stopped moving: he had changed into a piggy bank.

The cafe's customers all leant in to look at the piggy bank. As they did so, the Sun skipped out of the door and flew away. Straight away, everyone, including the children, crowded into the street, to watch him go... Within a few seconds, he had vanished from view.

That day turned out rather overcast, for the Sun was a little tired. But from the day after onwards, the

Pole Star shone in the sky once more, and the ships that set out for America mostly arrived in America.

As for the little pig, the Sun had been right to doubt that he would be free very soon. Naturally, customers often leave tips. Naturally, Papa Sayeed never forgets to drop the coins into the piggy bank's slot. But since the children come and shake them out again, I won't say every day, but maybe several times a day, there is reason to fear that the little pig may never again be entirely full up!

Afterword

Children understand everything—as everybody knows. If I knew that children would be the only ones reading this book, I would not even think of writing an afterword. But, alas, I'm afraid that these tales will be read as much by grown-ups as by younger people. So I feel I should provide a few explanations.

Rue Broca is not a street quite like any other street. If you look at a map of Paris, you will see—or think you see—that rue Pascal and rue Broca cross the boulevard de Port-Royal at right angles. If, confident in your map-reading, you were to take your car and drive down this boulevard, expecting then to turn into one or other of these two side streets, you might go back and forth a hundred times between the Observatory at one end of the boulevard and

Gobelins station at the other, but you would not find either of those two streets.

So, you will ask me: are rue Broca and rue Pascal made-up streets? Not at all! They do exist. And they do indeed run, in nearly straight lines, from boulevard Arago to rue Claude-Bernard. Therefore, they ought to cross the boulevard de Port-Royal.

The explanation of this anomaly is not to be found on your map, for the map can only show two dimensions. As in Einstein's world, at this spot, the surface of Paris curves and passes right over itself, so to speak. Forgive me for drawing on the jargon of science fiction, but really, there is no other way to say this: as with rue Pascal, rue Broca forms a dent, a hollow, a dive into three-dimensional sub-space.

Now, leave your car in its garage and return to the boulevard de Port-Royal, but this time on foot. Set out from Gobelins station and forge ahead, along whichever pavement you prefer. At a certain point, you will see that the row of houses that lines the boulevard has a gap in it. Instead of marching along beside shops or the wall of an apartment building as usual, you are walking alongside a space, a space fenced off by a railing to stop you falling into it. On the same pavement, not much farther along, you'll see the head of a staircase that appears to plunge deep into the entrails of the earth, like the steps that take you down to the Metro. Go down this staircase without fear. Once at the bottom, you are by no means underground; in fact, you will be in rue Pascal. Above your head, you'll see something that looks like a bridge. This bridge is the boulevard de Port-Royal, which you have just left behind.

A little farther along the boulevard, you will find another such staircase, like the first, but this one leading down to rue Broca.

This is bizarre, but it is true.

Now, let's ignore rue Pascal—it is too straight, too wide, too short also to harbour any mystery—and look at rue Broca alone.

This is a twisty street, narrow, crooked and sunken. By virtue of the spatial anomaly that I have described, although both its ends come out in Paris, the street itself is not quite part of Paris. No distance away, but on another plane, underground yet in the open air, this street by itself forms something like a small village. For the people who live there, this gives it a rather special feeling.

First, everybody knows everybody, and each one of them knows more or less what the others do and what they're busy with, which is exceptional in a city like Paris.

And then, the majority of them come from all kinds of different places; very few are from Paris. In this street, I have met Berbers, Algerian French, Spaniards, Portuguese, Italians, a Pole, a Russian... even a few French people from other parts of France!

Still, the people of rue Broca share at least one common pleasure: they love stories.

I have had many misfortunes in my literary career, the majority of which I attribute to the fact that the French in general—and Parisians in particular—do not like stories. They demand the truth, or, failing that, plausibility, realism. While the only stories that really interest me are those about which I am certain, from the start, that they have never happened, will never happen and could never happen. I feel that, due to the basic fact that it makes no documentary or ideological claims to justify its existence, an impossible tale has every chance of containing a good deal more profound truth in it than any story that is merely plausible.

Which perhaps makes me—I console myself—more of a realist in my own way than all those who claim to seek the truth, and who spend their lives stupidly ruled by insipid lies—lies that are indeed realistic purely by virtue of how insipid they are!

And now—one occasion does not make a habit!—here is a true story:

At number 69, rue Broca (I know, I know! I shall now be accused of God knows what dreadful innuendo. But what can I do? It was at number 69, not 67 or 71. For all you lovers of truth, this is one for you!). As I was saying, then: at number 69, rue Broca, there is a cafe-grocer's, the owner of which, Papa Sayeed, is a Berber married to a Breton woman. At the time of my story, they had four children: three girls and one boy (they had a fifth child later). The eldest girl is called Nadia, the second Malika, the third Rashida, and the little boy, who at the time was the youngest child, is called Bashir. Next to the cafe, there is a mansion house. In this house, among other tenants, lives a certain Monsieur Riccardi, Italian as his name suggests, also the father of four children, of whom the eldest is called Nicolas and the youngest is called Tina. I am leaving out other names, because there's no need for them and they would only be confusing.

Nicolas Riccardi often played in the street with the Sayeed children, for his father was a regular customer at the shop. This had been going on for a while and nobody would have dreamt of writing any of it down in a book had a certain odd character not one day turned up in the area.

This person was known as Monsieur Pierre. He was fairly tall, with chestnut hair that stuck up in spikes like a hedgehog, browny-green eyes and glasses. He always wore a two-day-old beard (people even wondered how he managed to keep his beard in what is usually a very temporary state, for a beard) and his clothes, such as they were, seemed always on the verge of falling apart. He was forty years old, a bachelor, and he lived up above on the boulevard de Port-Royal.

He came to rue Broca only to frequent the cafe, but he was often there and at all hours of the day. Besides, his tastes were modest: he appeared to live mainly on biscuits and chocolate, also on fruit when there was any, and all washed down with a great number of milky coffees and mint teas.

When he was asked what he did, he would reply that he was a writer. As his books were never seen anywhere, especially not in bookshops, this reply satisfied nobody, and for a long time the population of rue Broca wondered what he really did for a living.

When I say the population, I mean the grown-ups. The children never wondered anything of the sort, for they had understood right away: Monsieur Pierre was keeping his cards close; he was not a man like other men, really he was an old witch!

Sometimes, trying to unmask him, they would dance around him calling:

"Witch, old witch with your coconuts!"

Or again:

"Witch, old witch with your rubber jewellery!"

Instantly, Monsieur Pierre would throw off his disguise and become what he really was: he would wrap his old raincoat around his head, leaving only his face uncovered, let his thick glasses slide down to the end of his crooked nose and scowl frightfully. Then he would pounce on the kids, with all his claws out, giving a high, shrill, nasal cackle, something like the bleat of an old nanny goat.

The children would run away as if they were dreadfully afraid—but really they weren't as frightened as all that, for when the witch got hold of any of them, they would wriggle around and beat her off with their fists; and they were quite right to do so, for that is how we should treat old witches. They are only dangerous when we are afraid of them. Unmasked and shown who's in charge, they become rather good fun. At this stage, they can be tamed.

So it was with Monsieur Pierre. Once the children had forced him to reveal his true identity, everyone (starting with Monsieur Pierre himself) was greatly relieved, and normal relations were soon established.

One day when Monsieur Pierre was sitting at a table, enjoying one of his endless milky coffees, with the children clustered around, he began, of his own accord, to tell them a story. The next day, at their request, he told another one, and then on the days that followed, he told still more stories. The more he told, the more the children asked him to tell. Monsieur Pierre was obliged to start rereading all the collections of stories that he had ever read since his own childhood, simply in order to satisfy his audience. He told them stories from Charles Perrault, the tales of Hans Andersen and the Brothers Grimm, Russian stories, Greek, French and Arabic tales... and the children are still asking for them!

After a year and a half, having no more stories left to tell, Monsieur Pierre made the children a proposal: they would all meet every Thursday afternoon and together they would make up brand-new stories. And if they could come up with enough stories, the stories could be put into a book.

Which is what they did, and that is how this collection came about.

The stories in the collection were, thus, not written by Monsieur Pierre alone. They were improvised by him* in collaboration with his listeners—and whoever has not worked in this way may struggle to imagine all that the children could contribute, from solid ideas to poetic discoveries and even dramatic situations, often surprisingly bold ones.

---

* Apart from 'The Witch in the Broom Cupboard', which is inspired by Russian folklore.

I'll give a few examples, so first of all the first sentences in 'The Pair of Shoes':*

"There once was a pair of shoes that got married. The right shoe, which was the man, was called Nicolas, and the left shoe, which was the lady, was called Tina."

These few lines, which form the seed for the story that follows, come from young Nicolas Riccardi, whose little sister's name does indeed happen to be Tina.

Scoobidoo, the doll who knew everything, really existed, as did the guitar that became firm friends with the potato. And even as I write these words, the cunning little pig is still making himself useful as the piggy bank in Papa Sayeed's cafe.

On this same cafe's counter, in 1965, there was also a glass bowl that held two little fish, one red, the other yellow with black spots. It was Bashir who first realized that these fish could be "magic", and this is why they appear in 'The Witch in the Broom Cupboard'.

---

* Pierre Gripari originally wrote this afterword for a larger collection, which is why you may not recognize this story name or some of those that follow. If you want to read all the stories mentioned here, you can find them in *The Good Little Devil and Other Tales*!

As for those who will say that these stories are too serious for children, I offer the following reply in advance, with the help of one last example:

In an early version of the tale titled 'Uncle Pierre's House', my ghost only realized that he was a ghost thanks to the little girl amusing herself by putting her hand through his ethereal leg. It was Nadia, Papa Sayeed's eldest daughter, who had the inspired idea of having the little girl sit in the same armchair as the ghost, so that, on waking, the ghost discovers her sitting "right inside Uncle Pierre's tummy". These last few words are Nadia's own. Can grown-ups appreciate the symbolic value and moral beauty of this marvellous image? Our poor old ghost, a perfect specimen of the hardened, shrivelled-up, embittered old bachelor, is suddenly able to see himself as he really is. Suddenly liberty is within his reach, and truth, and generosity; he is—in short—set free, and his new freedom begins from the very moment when he symbolically becomes a mother. My friend Nietzsche also writes, I don't recall exactly where, of men as mothers... Yet it took a little girl to come up with this perfect idea!

But I'll stop here, for it would after all be a bit

much if, in a book intended for children, the afterword meant for the adults were itself to take up more space than your average fairy tale!

In any case, I haven't much else to add, except to wish my young friends from rue Broca happy reading, and the same to all who live on other streets in other towns, everywhere.

Pierre Gripari, 1966

Translator's Acknowledgements

I would like to thank Audrey Stanton and Etty Bo Tedman specially for their very helpful reading, also Michelle Stanton and Kerry Bell for their interpretation and comments, and Harold Lewis, who is always a trusty reader.

Sophie Lewis

Pierre Gripari, the Author

Pierre Gripari was born in 1925 in Paris, to a French mother and a Greek father. He studied at the prestigious Louis-le-Grand lycée, and tried his hand at various jobs, including serving in the army and acting as a trade-union delegate for an oil company.

He resigned in 1957 in order to become a writer, but it was not until the 1970s that he became famous, with the publication of his *Contes de la rue Broca*, translated in this volume. In these tales, the giants, witches and mermaids of traditional fairy tales leap from the page, animated by a very modern spirit. Blessed with a healthy disrespect for authority, the author took great pleasure in upsetting the natural order of the fantastic.

Pierre Gripari died in Paris in 1990.

Puig Rosado, the Illustrator

Puig Rosado was born in Spain on April Fools' Day, 1931—a date of birth that is surely not entirely free of responsibility for the course the rest of his life then took. His humorous posters, drawings and cartoons have been published in numerous countries, his work is on display in museums across Europe, and he has been honoured with many prizes. Puig Rosado is absolutely convinced that people with no sense of humour go, without exception, to hell!

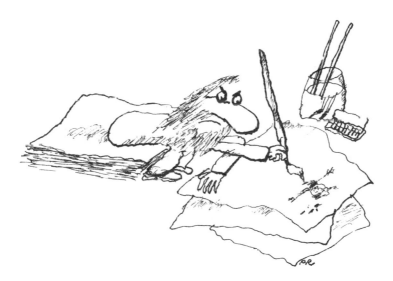

PUSHKIN CHILDREN'S BOOKS

Just as we all are, children are fascinated by stories. From the earliest age, we love to hear about monsters and heroes, romance and death, disaster and rescue, from every place and time.

In 2013, we created Pushkin Children's Books to share these tales from different languages and cultures with younger readers, and to open the door to the wide, colourful worlds these stories offer.

From picture books and adventure stories to fairy tales and classics, and from fifty-year-old bestsellers to current huge successes abroad, the books on the Pushkin Children's list reflect the very best stories from around the world, for our most discerning readers of all: children.